Praise for *The Search for*

"Richard Ellis has written the definitive giant sq____
scientific reporting and cultural history. If you are ____
natural mystery is all but conquered, this is the book ___ ___
— *Whole Earth Catalog*

"Ellis [has] the freedom and luxury to dredge into the depths of mythology, fantasy, fiction, and undocumented 'eyewitness' accounts, as well as into the scientific literature, itself not entirely free from imagination. The mix of all these sources, dissected and analyzed with an appropriate dash of skepticism, and an occasional pinch of speculation, yields a volume that surely will be read and re-read by the full range of readers from monster-lovers to scientists."
—Clyde Roger, Ph.D., National Museum of Natural History, Smithsonian Institute

"A gold mine of fact and fantasy, for we scientists who work on cephalopods and for all of us who love monsters."
—Martin Wells, *Nature*

"In researching an animal about which so little is known, Richard Ellis has managed to assemble an astonishing amount of material: history, mythology, biology, stories, models, and even a made-for-television movie. This is an amazing book about an amazing creature."
—Peter Benchley, author of *Jaws* and *Beast*

"An absorbing work . . . elegant, informative, passionate."
—*Publishers Weekly*

"Ellis has assembled a potpourri of ancient myth, rare sightings, and occasional bodily remains into a mosaic on *Architeuthis*, the *Kraken*, cryptic squids of the abyssal depths."
—David Bulloch, author of *Underwater Naturalist* and *The Wasted Ocean*

"A splendid job bringing together virtually every known account (mythical, fictional, and factual), and producing a narrative at once gripping and meticulously balanced."
—Oliver Sacks, M.D., author of *The Man Who Mistook His Wife for a Hat*

"He provides a huge amount of information, dealing not only with the scientific aspects of the giant squid but also with its scary appearances in film and fiction."
—*Scientific American*

"Ellis is fun to read, knowledgeable, and enthusiastic."
—*The Washington Post*

"Ellis's illustrated books about sea creatures have long made delectable reading. Here he explores the clammiest leviathan of the abyss, the tentacled, giant-eyed, giant squid . . . a fascinating compendium."
—*Booklist*

PENGUIN BOOKS

THE SEARCH FOR THE GIANT SQUID

Richard Ellis is one of America's most celebrated marine artists and writers. He is the author of nine books, including *Men and Whales*, *Great White Shark*, *Monsters of the Sea*, *Deep Atlantic*, and *Imagining Atlantis*. He lives in New York City.

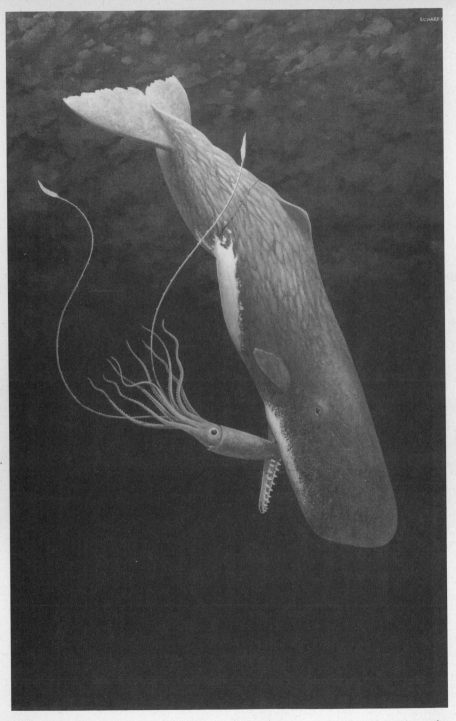

In the most fabled of deepwater encounters, a sperm whale captures a giant squid. Despite stories to the contrary, it is always the whale that initiates the contact, as the whale feeds on the squid and not vice versa.

Painting by Richard Ellis

The Search for the Giant Squid

Richard Ellis

PENGUIN BOOKS

PENGUIN BOOKS
Published by the Penguin Group
Penguin Putnam Inc., 375 Hudson Street,
New York, New York 10014, U.S.A.
Penguin Books Ltd, 27 Wrights Lane,
London W8 5TZ, England
Penguin Books Australia Ltd, Ringwood,
Victoria, Australia
Penguin Books Canada Ltd, 10 Alcorn Avenue,
Toronto, Ontario, Canada M4V 3B2
Penguin Books (N.Z.) Ltd, 182–190 Wairau Road,
Auckland 10, New Zealand

Penguin Books Ltd, Registered Offices:
Harmondsworth, Middlesex, England

First published in the United States of America by The Lyons Press 1998
Published in Penguin Books 1999

1   3   5   7   9   10   8   6   4   2

Portions of this book have previously appeared in *Monsters of the Sea*,
originally published by Alfred A. Knopf in 1995.
The chapter entitled "The Models of *Architeuthis*" appeared in *Curator* magazine
(Vol. 40, No.1, 1997) in a slightly modified form.

THE LIBRARY OF CONGRESS HAS CATALOGED THE HARDCOVER EDITION AS FOLLOWS:
Ellis, Richard, 1938–
The search for the giant squid/Richard Ellis.
p.   cm.
Includes bibliographical references and index.
ISBN 1-55821-689-8 (hc.)
ISBN 0 14 02.8676 4 (pbk.)
1. Giant squids.   I. Title.
QL430.3 A73E58   1998
594'.58—dc21   98–10436

Printed in the United States of America
Set in Adobe Garamond
Designed by Joel Friedlander, Marin Bookworks

Contents

A vast pulpy mass, furlongs in length and breadth, of a glancing cream-color, lay floating on the water, innumerable long arms radiating from its centre, and curling and twisting like a nest of anacondas, as if blindly to clutch at any hapless object within reach. No perceptible face or front did it have; no conceivable token of either sensation or instinct; but undulated there on the billows, an unearthly, formless, chance-like apparition of life.

As with a low sucking sound it slowly disappeared again, Starbuck still gazing at the agitated waters where it had sunk, with a wild voice exclaimed—"Almost rather I had seen Moby Dick and fought him, than to have seen thee, thou white ghost!"

"What was it, Sir?" said Flask.

"The great live squid, which, they say, few whale-ships ever beheld and returned to their ports to tell of it."

—*Moby-Dick,* 1851

The Search for the Giant Squid

Introducing
Architeuthis

A rchiteuthis—pronounced Ark-i-*tooth*-iss—is the scientific name of the giant squid. Actually, it is only the first half of the scientific name of the giant squid; the second half varies from species to species, assuming there are different species and not just one with geographical variations. In the chapter "By What Name Shall We Call the Giant Squid?," the possibility of different species is discussed, but since the name *Architeuthis* will be appearing regularly throughout this narrative, a proper introduction is necessary.

According to a system developed in 1758 by the Swedish naturalist Carl von Linné (known as Linnaeus), every living thing is to have two names: a *generic* name, which identifies the genus to which it belongs, and a *specific* name, which designates the species. Because they belong to the same genus, your dog and the wolf are, respectively, *Canis familiaris* and *C. lupus*. The coyote is *C. latrans*, but because it has different anatomical characteristics, the red fox has been placed in a separate genus, and it is known as *Vulpes vulpes*.

Often (but not always) the scientific names in Linnaeus's system of binomial nomenclature are derived from Greek or Latin words that are descriptive of the animal in question. The name *Architeuthis* can be broken down into its two components, *Archi* and *teuthis*. The first part is derived from the Greek *archae* or *archi*, meaning "first" or "early." The Greek word for squid is *teuthis*, so *Architeuthis* simply means "first squid." Since it was certainly not the earliest squid described, the name was probably meant to signify that it was first in importance or size.

Despite its spectacular appearance in movies like *Twenty Thousand Leagues Under the Sea* and *Beast*, we are not on familiar terms with the creature the ancients called the "kraken." Papier-mâché or fiberglass models in museums have given us a sense of its size and shape, but they have not captured its mystery and vitality. Only real-time film of the creature in its natural habitat can communicate the heretofore elusive reality of *Architeuthis*—the only living animal for which the term *sea monster* is truly applicable.

In January 1997, an expedition sponsored by National Geographic Television, the Smithsonian, the New England Aquarium, and several other institutions left for New Zealand, in hope of obtaining the first images of the living giant squid. The expedition leader was Clyde Roper, a Smithsonian Institution teuthologist (squid specialist), unquestionably the world's foremost authority on *Architeuthis*. He had assembled the expedition and raised the million dollars that it was going to cost. He had published any number of papers on this species, and whenever anyone wanted an opinion about the giant squid—how big it got, where it lived, what it ate—they would go to him. And in every one of these television, newspaper, or magazine interviews, he said that nobody has ever seen a living giant squid. "Then how do we know they exist?" was inevitably the interviewer's next question. "They have been washing up on beaches around the world for the last four hundred years," he would answer.

In addition to Roper, the group included a British expert on squids who was particularly knowledgeable about their relationship with sperm whales; a dedicated monster hunter from Bermuda; a National Geographic photographer; and a large cohort of support personnel. (Peter Benchley was originally scheduled to be the chronicler of this expedition, but he had to pull out at the last minute because of back problems.) They would be using two ships, on which they had loaded every piece of equipment that would contribute to the success of the mission: underwater cameras, robotic cameras, and cameras that could be affixed to sperm whales, in hope that a whale would act as cameraperson as it was chasing its dinner.

The expedition selected Kaikoura Canyon as their destination because it was a favorite haunt of large sperm whales, creatures known to feed on giant squid. The presence of the huge, ivory-toothed whales surely meant that *Architeuthis* was not far away. Roper and his team were buoyed by the fact that during the previous year, four giant squid had been trawled up by fishermen in New Zealand locations that seemed even less likely than Kaikoura Canyon to hold *Architeuthis*. Now these squid hunters were heading to what they believed was an ideal place to spot a living example of this traditionally elusive and enigmatic creature.

This was not the first such expedition; during the late 1980s, Frederick Aldrich, a Newfoundland scientist and Roper's predecessor as "world's foremost authority on the giant squid," had tried the same thing in a submersible with less

3

sophisticated camera equipment. Of course they wanted to get some pictures, but basically, Aldrich and Co. just wanted to *see* the damn thing. Roper and his team had much more ambitious plans. In its early stages, their expedition had also planned on a submersible, but they lacked the necessary time and money, so they'd decided to conduct their search by sending unmanned, remote-controlled cameras into the depths of the canyon, with the hope that a giant squid would pass in front of the lens.

No giant squid appeared, but the participants gathered data and experience that would be invaluable for future study, and would set the stage for further expeditions.

We know about a lot of other weird and mysterious creatures from the depths: fishes that light up, others that can swallow prey larger than themselves, still others that attach themselves permanently to members of the opposite sex so they don't have to search the inky blackness for a mate. We have obtained still photographs and even videos of other squids, some of them of a respectable size. A completely unexpected shark species, over 14 feet long and weighing 1,600 pounds, was hauled up from deep Hawaiian waters in 1976. Even more surprisingly, researchers in 1977 discovered a completely unexpected assemblage of living creatures—10-foot-long tube worms; giant, snow-white clams and crabs—that subsisted on hydrogen sulfide instead of oxygen.

So, how is it possible that an animal that can reach a length of 60 feet and weigh over a ton has remained out of sight for so long?

The giant squid is a real-life enigma. It differs dramatically from beasts like the unicorn, the dragon, and the sea serpent, all of which come equipped with a durable and honorable mythology but are lacking in corporeal evidence. Nor does it resemble creatures like the okapi, the coelacanth, and the megamouth shark, large animals whose existence was not suspected until a specimen fortuitously fell into the hands of a scientist.

Over the centuries, the verification of these creatures most often came from seafarers' tales, embellished by equal parts of ignorance and fear. Less frequently, the evidence washed up on the beach, but even when this happened, spectators were more than a little inclined to improvise and fantasize, particularly because many of these creatures had never been seen before and resembled nothing that landsmen had ever encountered, even in their most frightening nightmares. More recently, the carcasses of giant squid have been caught in deep nets, or even found floating at the surface.

Up to now, the giant squid has revealed its secrets most reluctantly. Virtually everything we know about this creature we have learned from the examination of dead or dying animals. No one has ever seen a living, healthy giant squid. From the sixteenth century to the present day, at scattered locations around the world, giant cephalopods, with tentacles 40 feet long and eyes the size of dinner plates, have washed ashore. On rare occasions, fishermen (most commonly in Newfoundland, New Zealand, and Norwegian waters) have encountered a living but moribund squid floating at the surface, but these creatures invariably died shortly after their discovery. (The very fact that these normally deep-water animals were at the surface indicated that they were already in trouble.) The most reliable descriptions of a living *Architeuthis* usually involve a sperm whale, and in most cases, the squid was on the way to becoming the whale's dinner. This leaves us with accounts coming from those who saw disembodied tentacles or disarmed bodies. Sperm whalers knew of the existence of giant cephalopods, because the great whales sometimes regurgitated large pieces of something just before they died, and the whalers often hooked the pieces and dragged them alongside. Although these pieces of squid were proof of the existence of a giant squid Melville wrote about them in *Moby-Dick* in 1851—it was not until 1861 that the first specimen was observed at sea.

Crew members of the French warship *Alecton,* sailing off the Canary Islands, spotted a large floating carcass and approached it (shooting at it as they came, just to be on the safe side) with the idea of bringing it aboard. They got a rope around the animal's 18-foot-long body, but as they were hauling it in, the body separated from the tail, and they were left with only the tail section. Nevertheless, they brought what they had to Tenerife, and the subsequent description, published in a French scientific journal, was the first description of the *poulpe géant* (giant octopus). In 1856, the Danish scientist Japetus Steenstrup, working from a beak and some three-hundred-year-old descriptions of "sea monsters," had named the giant squid *Architeuthis,* but it would be a number of years before people were convinced that a squid could grow so large.

Since the mid–nineteenth century a number of giant squid have been found in a few places around the globe. Most recently, however, fishermen have been pulling them up in unprecedented numbers off the coasts of New Zealand and the adjacent islands. In 1994, New Zealand scientists Gauldie, West, and Förch published a paper that contained the startling news that *"a total of 24 specimens of giant squid were recovered between 1983 and 1988 in New Zealand waters."* (My italics.) The largest was a female with a mantle length of 83.46 inches (just under 7 feet), and the smallest was another female whose mantle measured 36.27 inches.

In March 1995, off the coast of South Australia, the carcass of a recently dead 30-foot (total length) female was found floating on the surface. It was collected by a local fisherman, who brought it to the South Australian Museum in Adelaide for identification. The head and body measured 6 feet in length, the longer tentacle was 24 feet long, and the eye was 6 inches across.

Between January and August 1996, three large female giant squid were caught by Australian fishermen off the coast of Tasmania. One had sperm packets embedded in two of its arms. Exactly what this means in terms of the mating habits and reproductive biology of *Architeuthis* can only be guessed at. A total of four giant squid was caught in Australian waters in 1996, and four more off the shores of New Zealand. Where once the main focus of giant squid studies was centered in the North Atlantic, particularly Newfoundland and Norway, it has now shifted around the world to the Southern Ocean. Why? We haven't the faintest idea.

The giant squid is one of the largest animals in the world, but we do not know how big it actually gets. The maximum size of the giant squid has long been a subject for speculation among scientists, seamen, whalers, authors, and almost anyone else with an interest in the sea's larger inhabitants. The self-appointed authority on such matters, Gerald Wood's *The Guinness Book of Animal Facts and Feats,* tells us that "the largest squid so far recorded" was the Thimble Tickle specimen of 1878, which measured 55 feet from tail tip to tentacle tip. (Here it is worth repeating Arthur C. Clarke's remark that "it would be strange indeed if the world's biggest squid had been among the very few cast ashore to be examined and measured by naturalists. There may well be specimens more than a hundred feet in length.") On Clarke's 1988 video series *Mysterious World,* Frederick Aldrich examined a 20-foot-long immature specimen of *Architeuthis* and said, "I believe the giant squid reach an approximate maximum size of something like one hundred and fifty feet." It is difficult to imagine why Aldrich would have made such an irresponsible statement, unless it had to do with being on camera. (Since one cannot disprove a negative, one cannot say that such a length is impossible.) From the physical evidence, it would appear that the largest known specimen was the 57-footer that washed ashore in New Zealand in 1887.

Because *Architeuthis* is such a spectacular animal, those who would include it in their catalog of monsters often increase its length substantially, and often its weight as well. In a Time-Life book titled *Dangerous Sea Creatures,* for example, Thomas Dozier introduces his discussion of giant squid by saying that "two 42-foot tentacles were vomited by a captive whale in an aquarium, and experts calculated that these had to belong to a monster measuring at least 66 feet and weighing better than 85,000 pounds." Later, Dozier calls a 50-footer "ordinary," and says that there have been sperm whales captured with "tentacle marks 18 inches across, which would have to have been inflicted by a gargantuan squid of at least 200 feet long."

There is no evidence whatsoever that any squid can or does emit some sort of ammonium slime, as the killer squid does in Peter Benchley's *Beast.* In this novel, the "marine biologists" know it is *Architeuthis* that is killing all those innocent people because it leaves a powerful smell of ammonium wherever it goes, and also because it drops an occasional talon. There are indeed squids with claws, but *Architeuthis* is not one of them.

We don't know if *Architeuthis* is a solitary hunter, or if it congregates in schools, but the conventional wisdom holds that it is a loner. A single 60-foot-long giant squid represents the stuff of nightmares, with its writhing arms and saucer-sized eyes. The sperm whale at a maximum of 60 feet, the sei whale at 65, the fin whale at 80, and the blue whale at more than 100 are the only living animals that can surpass the giant squid in length.

Moreover, *Architeuthis* is assembled of a strange collection of parts that bespeak an alien creature rather than one we are used to. Even if we forgive *Architeuthis* its ominous size, it is still built according to some preternatural plan. There is nothing familiar or comforting about a spindle-shaped creature with a collection of writhing arms at one end and a pointed tail at the other. It seems to have no head—at least not where we expect to find one—but it does possess huge, lidless eyes; eight grasping arms studded with toothed suckers; elongated, whip-like tentacles with which to grasp its unfortunate prey; and a beak where no creature is supposed to have a beak—between its arms. It is the least-known large animal on earth, the last monster to be conquered.

Is the Sea Monster a Giant Squid?

Below the thunders of the upper deep,
Far, far beneath the abysmal sea,
His ancient, dreamless, uninvaded sleep
The Kraken sleepeth; faintest sunlights flee
Above his shadowy sides: above him swell
Huge sponges of millennial growth and height;
And far away into the sickly light,
From many a wondrous grot and secret cell
Unnumber'd and enormous polypi
Winnow with giant arms the slumbering green.
There hath he lain for ages and will lie
Battening upon huge seaworms in his sleep,
Until the latter fire shall heat the deep;
Then once by man and angels to be seen,
In roaring he shall rise and on the surface die.

   —Alfred, Lord Tennyson
    "The Kraken"
    Poems, Chiefly Lyrical, 1830

Architeuthis, the giant squid, is probably responsible for more myths, fables, fantasies, and fictions than all other marine monsters combined. In the *Odyssey* we read of Scylla, a horrible monster:

Her legs—and there are twelve—
are like great tentacles,
unjointed, and upon her serpent necks
are borne six heads like nightmares of ferocity
and triple serried rows of fangs and deep
gullets of black death. Half her length she sways
her heads in the air, outside her horrid cleft,

hunting the sea around that promontory
for dolphins, dogfish, or what bigger game
thundering Amphitrite feeds in thousands.
And no ship's company can claim
to have passed her without loss and grief; she takes
from every ship, one man for every gullet.

Also known as the kraken, the polyp (*poulpe* in the French of *Vingt mille lieues sous les mers*), and the sea serpent, the giant squid is, next to the shark, perhaps the most infamous animal in the sea. Aristotle introduced us to the *teuthos*, the giant squid, as differentiated from *teuthis*, the smaller variety. Somewhat later, in his *Naturalis historia*, Pliny discussed a "polyp" that plucked salted fish from the fish ponds of Carteia (on the Atlantic coast of Spain), and "brought on itself the wrath of the keepers, which owing to the persistence of the theft was beyond all bounds." The guards that surrounded the polyp were

> astounded by its strangeness: in the first place its size was unheard of and so was its color as well, and it was smeared with brine and had a terrible smell; who would have expected to find a polyp there, or who would recognize it in such circumstances? They felt they were pitted against something uncanny, for by its awful breath it also tormented the dogs, which it now scourged with the ends of its tentacles and now struck with its longer arms, which it used as clubs, and with difficulty they succeeded in dispatching it with a number of three-pronged harpoons.

The head was as big as a cask and held 90 gallons; its arms ("knotted like clubs") were 30 feet long with suckers like basins holding 3 gallons, and teeth corresponding to its size. Its remains weighed 700 pounds. The longer tentacles identify the polyp as a squid, rather than an octopus, but whatever it was, this would appear to be the only record of a cephalopod coming ashore for any reason except to die. (Pliny told us that it was "getting into the uncovered tanks from the open sea," so perhaps these were somehow accessible to a deep-water inhabitant. The fact that the tanks held "salted fish" indicates that they were not holding tanks at sea, however.) Pliny, of course, depended upon the reports of others for his *Naturalis historia*, but if we assume that something like this actually happened in antiquity (even allowing for exaggeration), it is the only such occurrence in all the literature.

The many-armed sea beast lay relatively dormant, however, until it was resurrected by Olaus Magnus (1490–1557), the Catholic archbishop of Sweden, a veritable fount of information on sea monsters. In his 1555 *Historia de gentibus septentrionalibus* (History of the People of the Northern Regions), he describes and illustrates several "monstrous fish," as follows:

> Their forms are horrible, their Heads square, all set with prickles, and they have long sharp horn round about like a Tree rooted up by the Roots: They are ten or twelve Cubits[*] long, very black, with huge eyes . . . the Apple of the Eye is of one Cubit, and is red and fiery coloured, which in the dark night appears to Fisher-men afar off under Waters, as a burning fire, having hairs like goose feathers, thick and long, like a beard hanging down; and the rest of the body, for the greatness of the head, which is square, is very small, not being above 14 or 15 Cubits long; one of these Sea-Monsters will drown easily many great ships provided with many strong Marriners.

Olaus Magnus's *Historia de gentibus septentrionalibus* appeared in English in 1658 as *A Compendious History of the Goths, Swedes and Vandals, and Other Northern Nations.* The descriptions and drawings that appeared on his maps firmly established the existence of many fabulous creatures, and were copied, reproduced, and modified for centuries, thus ensuring his place as one of the most important figures in the history of zoology. He described the *Soe Orm* as follows: "A very large Sea-Serpent of a length upwards of 200 feet and 20 feet in diameter which lives in rocks and in holes near the shore of Bergen."

One of the best-known drawings in Olaus Magnus's work is the one known as *Les marins monstres & terrestres, lelquez on trouve en beaucoup de lieux es parties septentrionales* (The Sea and Land Monsters That Are Found in Many Northern Places). This woodcut, supposedly the work of Hans Rudolph Manuel Deutsch (active in Switzerland around the middle of the sixteenth century), includes the *Soe Orm. Les marins monstres* then became the basis for Conrad Gesner's *Historia*

---

*A cubit is a linear measure based on the distance from the elbow to the tip of the middle finger, or from seventeen to twenty-two inches. Noah's Ark, therefore, at three hundred cubits, was about five hundred feet long.

Olaus Magnus's Soe Orm, 1555

animalium (1551–58), which is considered to be the basis of modern zoological classification. As Daniel Boorstin wrote in *The Discoverers,*

> His *Historia animalium,* following Aristotle's arrangement, supplied everything known, speculated, imagined, or reported about all known animals. Like Pliny, he provided an omnium-gatherum, but now added the miscellany that had accumulated in the intervening millennium and a half. A shade more critical than Pliny, he still did not deflate tall tales, as when he showed a sea serpent three hundred feet long.

One of Gesner's more interesting drawings shows a hydra: a creature with seven heads, a long scaly body, two clawed feet, and a curled-under tail. In the text, the hinder part of the head was said to resemble a Turk's cap, which is not shown in the drawing—unless the "crown" on each of the seven heads is a Turk's cap. It is difficult to equate this seven-headed image with any known animal, but if we assume that the artist was trying to portray something with which he was completely unfamiliar, perhaps from descriptions that changed over time, it is not impossible to see the "heads" as arms, and the body as that of a large cephalopod. The clawed feet look like an imaginary addition (they have six toes), and while

Looking at Conrad Gesner's 1555 drawing of a "monstrous serpent," with its seven heads, long scaly body, and curled-under tail, it is not difficult to imagine that it came from a description of a giant squid. (The feet, however, are a bit of a problem.)

we will never know what Gesner had in mind, his "monstrous serpent" could be one of the earliest depictions of a giant squid.

Of course Gesner included many of the creatures originally depicted by Olaus Magnus, which were then repeated, often without change, by the Renaissance encyclopedist Edward Topsell, whose *Historie of Foure-Footed Beastes* appeared in 1607.* Although Topsell's marine "beastes" have no feet whatsoever, they are recognizably the drawings of Olaus Magnus. Ulysses Aldrovandi (*De piscibus*, 1613) and John Jonstonus (*Historia naturalis*, 1649) followed with their encyclopedias,

---

*Topsell reproduces Gesner's drawing, identifying it as "The HYDRA, supposed to be killed by Hercules," and repeats Gesner's account of a carcass that had been brought from Turkey to Venice around the year A.D. 550. It was presented to the French king, who saw in the monster a "mutation of worldly affairs" that foretold an imminent cataclysm.

In 1734, Hans Egede, the Bishop of Greenland, reported this web-footed monster. Despite its somewhat improbable position, Henry Lee saw the giant squid as a likely explanation.

and they faithfully repeated Magnus's drawings and fanciful stories, including sea monsters with features of a giant squid.

Two centuries after Olaus Magnus, another ecclesiastic, the Danish missionary Hans Egede (who eventually became the bishop of Greenland), visited that icy island early in the eighteenth century, in hope of converting the natives to Christianity. Two settlements were established, the first in 1721 and the second in 1723. While the Greenland Eskimos proved to be unsusceptible to the faith that was being foisted on them, they were disastrously sensitive to the smallpox virus carried by one of the Danish missionaries, and most of them died. In Egede's story, *Det gamle Grønlands nye perlustration* (published in 1741), he recounts the following episode:

> As for other Sea Monsters . . . none of them have been seen by us, or any of
> our Time that ever I could hear, save that most dreadful Monster, that showed
> itself upon the Surface of the Water in the year 1734, off our colony in 64
> degrees. The Monster was of so huge a Size, that coming out of the Water its
> Head reached as high as the Mast-Head; its Body was as bulky as the Ship, and
> three or four times as long. It had a long pointed Snout, and spouted like a
> Whale-Fish; great broad Paws, and the Body seemed covered with shell-work,
> its skin very rugged and uneven. The under Part of its Body was shaped like an

In Sea Monsters Unmasked, *Henry Lee suggested that this might be the explanation for Bishop Egede's sea serpent.*

enormous huge Serpent, and when it dived again under Water, it plunged backwards into the Sea and so raised its Tail aloft, which seemed a whole Ship's Length distant from the bulkiest part of its Body.

By this time the Dutch and the British were energetically slaughtering the bowheads of Greenland and Baffin Island for their baleen plates and oil, so Egede must have been familiar with whales. He had Pastor Bing draw a picture of the monster, which was reproduced in his *Perlustration* (published in English as *A Description of Greenland* in 1745), and since Egede was known to be a sober, reliable observer, the picture thus became one of the earliest illustrations of a sea monster based on a reliable eyewitness account. Does Bing's drawing look like a giant squid? Not really, but if we factor in the modifications attributable to ignorance and exaggeration, we are left with no other choice.

Clergymen seemed to have a particular affinity for sea monsters (or perhaps it was vice versa), for the next to write about the monstrous creature is Bishop Erik Ludvigsen Pontoppidan of Bergen, author of *The Natural History of Norway,* published in 1755. The good bishop firmly believed in the kraken, and maintained that the fishermen he spoke to told him it was a mile and a half in circumference. Referring to this beast as "the largest and most surprising of all the animal creation" and "incontestibly the largest Sea-monster in the world," the bishop wrote:

> It is called *Kraken** or *Kraxen*, or, as some name it, *Krabben* . . . He shows himself sufficiently, although his whole body does not appear, which in all likelihood no human eye ever beheld (excepting the young of this species) its back or upper part, which seems to be in appearance an English mile and a half in circumference, (some say more, but I chuse the least for greater certainty) looks at first like a number of small islands, surrounded with something that floats and fluctuates like sea weeds. . . . At last several bright points or horns appear, which grow thicker and thicker the higher they rise above the surface of the water, and sometimes they stand as high and large as the masts of middle-siz'd vessels.

By all accounts, what the bishop was describing was a giant squid.† "The creature," he added, "does not, like the eel or the land-snake, taper gradually to a point, but the body, which looks to be as big as two hogs-heads, grows remarkably small at once just where the tail begins. The head in all kinds has a high and broad forehead, but in some a pointed snout, though in others it is flat, like that of a cow or a horse, with large nostrils, and several stiff hairs standing out on each side like whiskers."

The nostrils and whiskers are somewhat problematical, but the rest of the portrayal is a remarkably accurate description of *Architeuthis*—by somebody who has no idea what sort of creature he is looking at. For landlubbers used to animals

---

*The Norwegian word *kraken* is popularly believed to be derived from a word that means "uprooted tree," from the similarity of the body and arms of a giant squid to the trunk and roots of a tree, but Jan Haugum, a Norwegian biologist and linguist, has explained to me that the old Norwegian word made its first appearance in Pontoppidan's 1755 work, and was used to mean nothing more or less than a "sea monster." *Kraken*, by the way, is the plural; the singular is *krake*. A *krakenbanke* was a reef or bank where the water was shallower than it was supposed to be, because there was a huge *krake* lying on the bottom. According to Haugum, *krabben* is a complete misnomer, since it simply means "crab."

†In the chapter of *Moby-Dick* called "Squid," Herman Melville makes reference to the good bishop: "There seems to be some ground to imagine that the great Kraken of Bishop Pontoppidan may ultimately resolve itself into Squid. The manner in which the Bishop describes it, as alternately rising and sinking, with some other particulars he narrates, in all this the two correspond. But much abatement is necessary with respect to the incredible bulk he assigns it."

with a head at one end and a tail at the other, the kraken's tail, with its pointed apex, would have been the head, and the arms trailing behind obviously suggested a tail. Pontoppidan continues: "They add that the eyes of this creature are very large, and of a blue color, and look like a couple of bright pewter plates. The

In his Natural History of Norway, *published in 1755, Bishop Erik Pontoppidan described this snake-like, maned sea serpent.*

whole animal is of a dark brown color, but it is speckled and variegated with light streaks and spots that shine like tortoise shell." He mentions two places, Amunds Vaagen in Nordfjord and the island of Karmen, where carcasses had been found at high tide.

Bishop Pontoppidan took the deposition of a certain Captain von Ferry, who claimed to have seen a "sea-snake" passing his ship in August 1746. Upon being informed of the presence of the serpent, von Ferry came about in order to get nearer to it. At the bishop's request, he described it in a letter written to the Court of Justice at Bergen:

> The head of this sea-serpent, which it held more than two feet above the water, resembled that of a horse. It was of a greyish color, and the mouth was quite black and very large. It had large black eyes, and a long white mane, which hung down to the surface of the water. Besides the head and neck, we saw seven

or eight folds, or coils, of this snake, which were very thick, and as far as we could tell, there was a fathom's distance between each fold.

Although this description doesn't immediately suggest one, von Ferry's serpent, with the "seven or eight folds . . . which were very thick," looks more and more like a squid. If one is familiar with snakes and has never seen (or expected to see) a giant squid, the long tentacles, which have been measured at 35 feet, suggest nothing more than some sort of a serpent. It is therefore possible that giant squid are not only responsible for sea-serpent sightings but are responsible for the very name.

An Englishman named Charles Douglas, sailing off Lapland in the HMS *Emerald* in 1769, on one of the first European oceanographic voyages, interrogated the Norwegians about the kraken and the sea serpent, and while no one could tell him anything about the kraken, they knew a lot about what they called "Stoor Worms." Douglas recounted one sighting, described by the master of a Norwegian vessel, of three of the worms, "floating upon the surface of the sea, twelve parts of the back of the largest appearing above water; each part being in length about six feet . . . so that upon the whole he judged the animal could not be less than twenty-five fathoms long, and about one in thickness." From Douglas's description (which was read to the Royal Society in 1770), the "worms" might very well have been the arms of a giant squid, either dead or dying, given what we now know about the presence of *Architeuthis* in Norwegian waters.

In the *Pictorial National Library* magazine for 1849, in an anonymous article about sea serpents, there is an account that, the author states, "will suffice to show that the sea-serpent in no stranger in the waters of the northern and eastern hemispheres." The description is attributed to the Reverend Mr. Deinboll, the archdeacon of Molde:

On the 28th of July, 1845, four men were out on the Ramsdale-fiord, fishing. About seven o'clock in the evening, a long marine animal was seen slowly moving forward. The visible part of the body appeared to be forty to fifty feet in length, and moved in undulations like a snake. Body round, dark color, and several feet thick. Its fore part ended in a sharp snout, and its immense head was raised above the water in a semi-circular form. The color of the head was dark

One of the illustrations that accompanied the original report of a "sea serpent" sighted from the ship Daedalus *in 1848 off the Cape of Good Hope. It is probably easier to see this creature as a giant squid—with its tail above the surface—than as a sea serpent.*

brown, the skin smooth. No eyes were noticed, nor mane nor bristles on the throat.

A hundred and one years later, in Vike Bay in the very same Romsdalfjord, a giant squid measuring 30 feet in total length was found by fishermen. It is hard to fault those who believed that the 1845 animal was a sea serpent. In fact, in describing the 1946 incident, Bjorn Myklebust wrote that before the squid stranded, it had been seen swimming around the fjord. "It is therefore possible," he wrote, "that the observed sea serpents may simply have been . . . giant squids that were lying and splashing at the surface. . . . It thus seems likely that many stories about sea serpents can be reduced to stories about giant squids."

In 1848, the frigate *Daedalus* was sailing off the Cape of Good Hope when the crew spotted

> an enormous serpent, with head and shoulders kept about four feet constantly above the surface of the sea, and as nearly as we could approximate by comparing with the length of what our main topsail yard would show in the water, there was at the very least 60 feet of the animal a *fleur d'eau,* no portion

of which was, to our perception, used in propelling it through the water, either by vertical or horizontal undulation.

This description came from Peter M'Quhae, the *Daedalus*'s captain, who was responding angrily to a request from the Admiralty that he confirm or deny the rumors about a sea serpent. M'Quhae went on to write that its diameter was "15 or 16 inches behind the head, which was, without any doubt, that of a snake," and that "it had no fins, but something like the mane of a horse, or rather a bunch of seaweed, washed about its back." The *Illustrated London News* ran the story, accompanied by drawings done from M'Quhae's description, and the *Daedalus* monster entered the lists as one of the best-described serpents of all time. Sir Richard Owen—who so disbelieved Darwin that he tried to have him excommunicated—attempted to argue that the animal was really a gigantic seal.

To his dying day, Sir Richard Owen refused to admit even the slightest possibility that monsters might exist outside the zoological framework that he understood. In response to one "sighting," Owen wrote, "The observers have no expert knowledge of zoology; their observation is, therefore, without merit." Of M'Quhae's sighting, he noted, "It is very probable that no one on board the *Daedalus* ever before beheld a gigantic seal swimming freely in the open ocean." With his penchant for bombastic overstatement, Owen finished off his critique (which was published in the *Times*) with the statement, "A larger body of evidence from eye-witnesses might be got together in proof of ghosts than of the sea-serpent." In his angry response (also printed in the *Times*), M'Quhae accused Owen of flagrantly misquoting him, and wrote:

> Finally, I deny the existence of excitement, or the possibility of optical illu-
> sion. I adhere to the statement, as to form, colour, and dimensions, contained
> in my official report to the Admiralty; and I leave them as data whereupon the
> learned and scientific may exercise the "pleasures of imagination" until some
> more fortunate opportunity shall occur of making a closer acquaintance with
> the "great unknown"—in the present instance assuredly no ghost.

If we look at the drawing of the *Daedalus* monster with *Architeuthis* in mind, the identification of the "monster" leaps off the page. We see not an "enormous serpent," but rather the tail of an enormous cephalopod. Captain M'Quhae

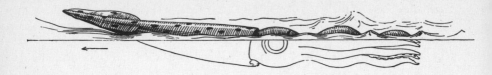

A drawing of a giant squid swimming at the surface—a possible explanation for the Daedalus's sea monster. The long tentacles have been eliminated from the drawing, but they would more than double its length.

obligingly drew an eye where he thought it ought to go, but in his description he never mentions it. He does, however, mention that "no portion [of its anatomy] was used in propelling it through the water, either by horizontal or vertical undulation." No vertebrate can move through the water without some visible means of propulsion, but a squid, which uses water ejected from the funnel or the mantle to move, could easily conform to M'Quhae's description.

HMS *Plumper* was in the North Atlantic in 1848 when a strange creature was sighted. In the *Illustrated London News* for April 10, 1849, an article, which was signed "A Naval Officer," contained the following description:

> Being due west of Oporto [Portugal] I saw a long black creature with a sharp head, moving slowly. I should think about two knots, through the water in a north westerly direction, there being a fresh breeze at the time, and some sea on. I could not ascertain its exact length, but its back was about twenty feet if not more above the water; and its head, as near as I could judge, from six to eight . . . the creature moved across our wake towards a merchant ship barque on our lee quarter, and on the port tack.

With a dark "body," pointed head, no mouth, and only the faintest suggestion of an eye, the illustration that accompanied the description of the *Plumper's* sighting looks like no living creature—except possibly a giant squid.

I do not propose to assign all sea-serpent sightings to *Architeuthis,* but there is a strong possibility that many of those seamen and passengers who saw a "monster" of some sort were actually seeing the giant squid. Let us examine the serpents of the Scandinavian ecclesiastics Bishops Egede and Pontoppidan. As Henry Lee points out in *Sea Monsters Unmasked,* Egede's illustration of a sea serpent can be easily made to look like a giant squid if we eliminate the eye and the mouth and have the "blow" come from the funnel instead of from the mouth. Pontoppidan's accounts are based largely on hearsay and the report of von Ferry, as well as one from a Governor Benstrup, but he also mentions the carcasses that washed ashore with regularity on the coasts of Norway. Is it not possible that mid-eighteenth-century Norway saw an invasion of giant squid like the one that occurred in Newfoundland in the late nineteenth century and New Zealand most recently?

Even a limited review of the "sea serpent" literature produces any number of cases where the giant squid becomes a strong candidate. When British naturalist Philip Henry Gosse wrote *The Romance of Natural History* in 1861, he concluded with a chapter entitled "The Great Unknown," in which he summarized much of what was known about sea monsters up to that time, including a retelling of the 1845 kraken sighting in Romsdalfjord.

Gosse then quotes Sir Richard Owen, who wrote, "Few seacoasts have been more sedulously searched, or by more acute naturalists . . . than those of Norway. Krakens and sea-serpents ought to have been living and dying thereabouts from long before Pontoppidan's time to our day, if all tales were true; yet they have never vouchsafed a single fragment of the skeleton to any Scandinavian collector." Eleven years later, a kraken "fragment"—consisting of an entire giant squid—would appear off Newfoundland, and now that an explanation appeared, it became possible to interpret Gosse's story as a description of what the Norwegians now call *kjempeblekksprut*. Particularly significant are the "waving motion" behind the animal (the arms); the "head" (a tentacular club); and the absence of eyes.

In 1886, an Australian named Charles Gould wrote a book called *Mythical Monsters*. It is largely about Chinese and Japanese dragons and other Asian monsters (Gould traveled widely in the Far East), but there is a chapter devoted to sea serpents. It commences with a discussion of the ecclesiastics Bishops Pontoppidan and Egede, as well as the early annalists such as Olaus Magnus and Aldrovandi, and then gets to the meat of his discussion: accounts of unidentifiable animals at sea. He introduces one Arthur de Capell Brooke, who sailed around the northern coast of Norway in 1820, and published his observations in *Travels Through Sweden, Norway and Finnmark* in 1823. Brooke seemed to be fascinated by sea serpents, because he recorded many sightings:

Item: Off Otersoen. . . . It was of considerable length, and longer than it appeared, as it lay in large coils above the water to the height of many feet. Its head was shaped like that of a serpent; but he could not tell whether it had teeth or not. He said it emitted a very strong odour and that the boatmen were afraid to approach near it.

Item: At Alstahoeg I found the Bishop of the Nordlands [another clergyman!] who was an eye-witness to the appearance of two in the Bay of Sörsund, on the Drontheim Fjord. . . . They were swimming in large folds, part of which were seen above the water, and the length of the largest he judged to be about one hundred feet. They were of a darkish grey colour, the heads hardly discernible, from their being almost under water.

Captain Brooke believed these reports, not only since they came from the mouths of fishermen ("an honest and artless class of men, who, having no motive for misrepresentation, cannot be suspected of a wish to deceive"), but also because some of the informants were "of superior rank and education," including the governor of Finnmark, Mr. Steen; the clergyman of Carlsö, the Reverend Deinboll of Vadsö; and the bishop of Nordland.

Gould also included an account from a Mr. McLean, the parish minister of Eigg. In 1809, he was rowing along the coast of Coll (in the Hebrides) when he spied an animal just offshore: "Within a few yards of it, finding the shallow water, it raised its monstrous head above water and, by a winding course, got, with apparent difficulty, clear of the creek where our boat lay. . . . Its head was somewhat broad, and of a form somewhat oval, its neck somewhat smaller; its shoulders, if I can so term them, considerably broader, and it tapered towards the tail, which last it kept pretty low in the water." Given such scanty details, it is difficult to envision the animal, but with no mouth, eyes, ears, or mane, it bears little resemblance to any known quadruped. Again it is possible to refer the description to some sort of invertebrate, but admittedly, only in the vaguest terms. We must assume that McLean saw something that did not fit the description of any creature known at that time.

In *The Zoologist* (subtitled *A Popular Miscellany of Natural History*), another Hebridean sighting, this one in 1872, is presented in some detail in an essay by two gentlemen of the cloth, Messrs. John Macrae and David Twopeny. (Macrae was the minister of Glenelg, Invernesshire; Twopeny was the vicar of Stockbury, Kent.) Sailing down the Sound of Sleat, with "two ladies, F. and K., a gentlemen, G.B., and a Highland lad," the party beheld a dark mass about two hundred yards astern, which proceeded to rise in symmetrical lumps, then sank out of sight, and surfaced again. They estimated its total length as 45 feet. It then "began to approach us rapidly, causing a great agitation in the sea. Nearly the whole of the body, if not all of it, had now disappeared, and the head advanced at at a great rate in the midst of a shower of fine spray, which was evidently raised in some way by the quick movement of the animal—it did not appear how—and not by spouting."

The next day they spotted the creature again, only this time "it looked considerably longer than it did the day before . . . it looked at least sixty feet in length. Soon it began careering about, showing but a small part of itself, as on the day before, and appeared to be going up Lochourn." According to Macrae and Twopeny, several other people saw the creature. It was not until the book's publication that the authors were made aware of "Pontoppidan's Natural History or his print of the Norwegian sea-serpent, which has a most striking resemblance to the first of our own sketches." (Their essay is entitled "Appearance of an Animal, Believed to Be That Which Is Called the Norwegian Sea Serpent, on the Western Coast of Scotland, in August, 1872.")

It is worth noting that many "sea serpent" appearances that were recorded before *Architeuthis* was first described occurred in those waters later revealed to be the sometime haunts of the giant squid. Olaus Magnus said that the *Soe Orm* lived near the shore of Bergen; Bishops Pontoppidan and Egede saw their monsters off Norway and Greenland; the Reverend Dienboll's appeared in Romsdalfjord; and the controversial serpent of the *Daedalus* was spotted off the Cape of Good Hope. It seems more than a passing coincidence that sea serpents often inhabit those waters known to be occasionally occupied by giant squid, and vice versa.

In 1905, as if to dispel Owen's criticism that it was never scientists who saw these monsters, two naturalists, E. G. B. Meade-Waldo and M. J. Nicoll, on a scientific cruise aboard the Earl of Crawford's steam auxiliary yacht *Valhalla*, sighted a dark brown animal with a great frill on its back off the coast of Brazil.

In 1905, when Michael Nicoll, a naturalist aboard the yacht Valhalla, *sighted a sea serpent off Brazil, he drew this illustration, which was reproduced in the* Proceedings of the Zoological Society of London.

In the *Proceedings of the Zoological Society of London,* Meade-Waldo published his observations:

> I looked and saw a large fin or frill sticking out of the water, dark seaweed-brown in colour, somewhat crinkled on the edge. It was apparently about 6 feet in length and projected from 18 inches to 2 feet from the water. I could see, under the water to the rear of the frill, the shade of a considerable body. I got my field glasses on to it (a powerful pair of Goerz Trieder), and almost as soon as I had them on the frill, a great head and neck rose out of the water in front of the frill; the neck did not touch the frill in the water, but came out of the water in front of it, at a distance of not less than 18 inches, probably more. The neck appeared about the thickness of a slight man's body, and from 7 to 8 feet was out of the water. . . . The head had a turtle-like appearance, as had also the eye. I could see the line of the mouth, but we were sailing pretty fast, and quickly drew away from the object, which was going very slowly. It moved its head from side to side in a peculiar manner; the colour of the head and neck was dark brown above, and whitish below—almost white, I think.

Michael Nicoll's similar description was published alongside that of Meade-Waldo, and Nicoll also contributed a drawing, which is reproduced here. It is certainly possible to read their descriptions as a monster with a "frill," but it is almost as easy to see a description of a giant squid swimming at the surface: the

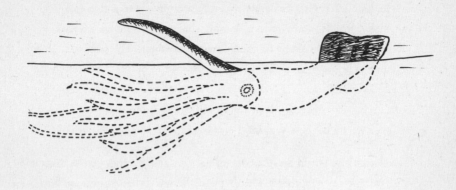

The Nicoll drawing adapted to conform to the giant squid hypothesis.

frill could be one of the tail fins, the neck (not attached to the "frill") one of the long tentacles, and the head its flattened end. The "turtle-like" appearance of the eye is admittedly a problem, but if something looks like the "head" of an animal, we expect to see an eye *somewhere,* and Meade-Waldo and Nicoll may have only imagined it. (There have been many "sea monster" sightings that can be explained as *Architeuthis* if we assume that the observers misidentified the eye.) As for the "peculiar" movement of the head (which Nicoll described as "a curious wriggling movement"), it is more likely that they were describing an object that was *not* a head, but something else—a tentacle, perhaps. We know virtually nothing about the locomotion of giant squid at the surface (if indeed they ever locomote at the surface), but there may be a possibility that these creatures swim with a tentacle out of the water, which would go a long way toward explaining the sightings of a large-bodied animal with a long neck. (If a giant squid ever swam with *both* tentacles out of the water, we would probably have reports of two-headed monsters.)

Major General H. C. Merriam of the U.S. Army sighted a sea serpent while sailing off Wood Island, Maine, in August 1905, and wrote about it in a letter to Dr. F. A. Lucas, director of the American Museum of Natural History in New York. The letter (reproduced in its entirety in the appendix to V. C. Heilner's *Salt*

Water Fishing) contains this description of a "monster serpent":

> Its head was several feet above the surface of the water, and its long body
> was plainly visible, slowly moving toward our boat by sinuous or snake-like
> motion. . . . It had no dorsal fin unless it was continuous. The color of its back
> appeared to be brown and mottled, shading down to a dull yellow on the belly.
> The head was like that of a snake, and the part shown above the surface—that
> is the neck—appeared to be about 15 to 18 inches in diameter. If it had any pec-
> toral fins we did not observe them. I estimated its length at 60 feet or more.

The Australian David Stead wrote about fish and fishing, whales and whaling, and, occasionally, monsters. He obviously believed that Australians had been
shortchanged in the sea-serpent department, so in *Giants and Pygmies of the Deep*
he recounts several Australian sightings. The first occurred off Bellambi Reef,
New South Wales, in 1930, and observers described a "vast monster of the serpent type," with a mouth like a pelican's beak. It is obvious to Stead that the open-
mouthed beast could only have been a rorqual whale, probably a minke. Only
two days, later, however, another monster appeared off Scarborough ("a few miles
north of the first occurrence"), but this time, it could not be so easily explained
away. It looked like a huge black snake, some 80 or 90 feet long, and it threw its
huge head in the air, then ducked below the surface, giving the impression that
it was feeding. Stead concludes that this really was a monster, "a great Calamary
or Cuttlefish or giant Sea Squid, of the type frequently called *Polypus*." He then
discusses several additional Australian sightings, and concludes by saying, "In all
of these and in many others there seems to me to be enough evidence to identify
the beasts seen, not as any kind of real serpent, but the terrible gigantic Calamary."

Then we have the tale of the Grace Line steamer *Santa Clara,* sailing from
New York to Cartagena in 1947. Some 118 miles off North Carolina, the ship
struck a marine creature. Third Officer John Axelson saw

> a snakelike head rear out of the sea about 30 feet off the starboard bow of
> the vessel. His exclamation of amazement directed the attention of the other two
> mates to the Sea Monster, and the three watched it unbelievingly as in a
> moment's time it came abeam of the bridge where they stood and was left astern.
> The creature's head appeared to be about 2½' across, 2' thick, and 5' long. The

cylindrically shaped body was about 3' thick, and the neck 1½' in diameter. . . .
The visible part of the body was about 35' long.*

Since none of the descriptions include a mouth or eyes—features that would probably attract immediate attention—it seems possible to assign all the elements of the *Santa Clara* sighting to the giant squid. The "snakelike head" could be the club end of one of the long tentacles, with the suckers turned away from the viewer, and the "neck" would be the tentacle itself. The given dimensions coincide with those of the club of a large squid's tentacle, and even the "cylindrically shaped body" is the right size. In other words, there is nothing in this description that might *not* refer to *Architeuthis.*

On October 31, 1983, road workers near Stinson Beach, north of San Francisco, spotted an "unidentified animal" swimming offshore. As reported in the *San Francisco Chronicle,* the animal was being followed by a flock of birds and about two dozen sea lions. One of the workers interviewed said, "There were three bends, like humps, and they rose straight up. Then the head came up to look around." On November 2, surfers reported a "sea serpent" near Costa Mesa, which they described as resembling "a long black eel." One of the surfers (quoted in the Costa Mesa *Daily Pilot*) said, "There were no dorsal fins. The skin texture wasn't the same as a whale, and when it broke water, it wasn't like a whale at all. I didn't see the head or the tail." From this sketchy information (first reported by J. R. Greenwell in the International Society for Cryptozoology's *Newsletter,* and then in a Time-Life book called *Mysterious Creatures*), it is difficult to form an opinion about the nature of this beast—assuming it was the same in both cases—but it is not unreasonable to visualize the elements that were described as being parts of a large squid. It must be pointed out, however, that there has never been a record of *Architeuthis* washing ashore in California. There is an animal known as the Pacific giant squid *(Moroteuthis robustus),* which does not get as large as *Architeuthis,* but can achieve a body and tentacle length of 19 feet (see pages 143–145).

In the conclusion to *In the Wake of the Sea-Serpents,* Bernard Heuvelmans wrote,

The legend of the Great Sea-Serpent, then, has arisen by degrees from chance

---

*The story appears in Bernard Heuvelmans's *In the Wake of the Sea-Serpents,* with no attribution except that a "detailed communique . . . was put out by the Associated Press."

sightings of a series of large sea-animals that are serpentiform in some respect. Some, like the oarfish, the whale-shark and Steller's sea-cow, have been unmasked in the last few centuries. But most remain unknown to science, yet can be defined with some degree of exactitude, depending on the number and precision of the descriptions that refer to them.

He does not mention the giant squid in this passage, but 30-foot tentacles are surely more "serpentiform" than any parts of a whale shark or Steller's sea cow. (He devotes a chapter to "The *Kraken* and the Giant Squid," but it serves more to introduce the creature and its colorful history than to explain its sea-serpent possibilities.) The giant squid does not explain everything; there are many sightings by reputable persons that cannot be conveniently dumped into a basket woven of squid tentacles.

Even if we assume that the aquatic dinosaurs are gone, and that there are no giant eels, giant snakes, giant seals, or giant otters, we are still left with some "sea serpent" sightings that are difficult to explain. It is perfectly natural for untrained persons to attribute "monster" characteristics to strange creatures seen at sea. In the past, of course, there were no Disney movies, and no *Nature* programs on PBS; people had to depend upon published descriptions, which, as we have seen, were often wildly exaggerated. Because hardly anybody has ever seen a giant squid, how could they possibly reconcile some strange, many-armed commotion with what David Stead called the "terrible gigantic Calamary"?

The impossibility of proving a negative proposition encourages all true believers to persevere; as long as it cannot be demonstrated that the monsters do *not* exist—and of course it cannot—they will continue to hope. More than any other creature, real or imaginary, the giant squid fuels the cryptozoological fires. Tales of ship-grabbing monsters are as old as seafaring, but it was not until some still unexplained oceanographic anomaly in the late nineteenth century that specimens of *Architeuthis* began appearing on the rocky beaches of Newfoundland. Because the giant squid is such an incredible animal, it is actually easier to assume that it is a mythological creature rather than a real one. Its melodramatic appearance in novels such as *Twenty Thousand Leagues Under the Sea* and *Beast* has supported its delegation to the world of fiction; like the white whale, *Architeuthis* seems too big and dangerous to be true.

The Biology
of Squids,
Giant and
Otherwise

In their 1996 *Cephalopod Behaviour,* Roger Hanlon and John Messenger wrote, "Cephalopods are among the most beautiful of all animals, and their behaviour is complex and fascinating." "Complex" is an understatement, and "fascinating" does not begin to do them justice. The name *cephalopod* comes from the Greek for "head-footed" and refers to the arrangement whereby the arms appear to spring from the head. All cephalopods have at least eight arms—octopuses quit there—and the cuttlefishes and squids have two additional tentacles that they can shoot out to capture prey. (The chambered nautilus, believed to be the most primitive of all living cephalopods, can have as many as ninety arms.) All squids have ten appendages, which classifies them as decapods, but the family Octopoteuthidae (eight-armed squid) has two feeding tentacles that are resorbed into the body as the animal matures, and the adults have only eight arms.

There are some seven hundred species of cephalopods—squids, octopuses, cuttlefishes, chambered nautiluses—and within this broad classification, there are some forty genera of squids.* They come in a dazzling array of sizes and shapes, equipped with claws, hooks, suckers, giant axons, eyes as complex as those of the "higher" vertebrates, beaks like parrots, or lights all over them. Some are tough and muscular, while others are as soft and gelatinous as jellyfish. Some species can fly, others can descend to abyssal depths; some live in uncountable congregations, while others are solitary hunters, prowling the abyssal depths in search of food. They are the most numerous and varied of all the cephalopods, officially classi-

---

*"It should be noted," wrote W. C. Summers, "that squid is both singular and plural. The language allows gatherers (hunters and fishermen) the option to refer to any number of the same sort by the singular term—squid. Where one species is concerned and there is a large-scale operation, we have no trouble with the exception—squid is plural. But the option is based on the perception of the reporter and how narrow a distinction that person wishes to make. *Squids* implies a collection of more than one kind, though it should be considered that the apparent plural may refer to species, numbers of one species, or even sizes (if these have a different market value). Scientific literature in English will usually reserve the plural term squids for multiple species; both words are used for several examples of the same species."

fied as mollusks, even though their shell (known as the *gladius* or "pen") is found inside the body. In the smaller species, the pen is less than a quarter of an inch long, while in *Architeuthis*, it can be as long as the mantle, as much as 8 feet in length. (The familiar chalky material used in birdcages is the pen of the cuttle-fish, known as the cuttlebone.) The size range for the various squids is enormous, from the tiny *Pickfordiateuthis* (approximately the length of its name printed here), to the subject of this book, the largest of all invertebrates.

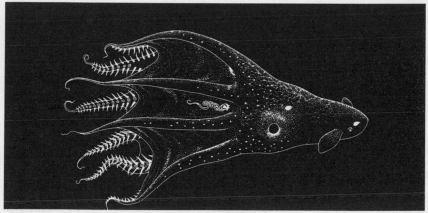

The "vampire squid from Hell," Vampyroteuthis infernalis is neither squid nor octopus. It is a deep-water cephalopod that has recently been observed to move quickly, rather than passively hovering. It reaches a maximum length of 12 inches.

The deep-water cephalopod *Vampyroteuthis infernalis* appears to be neither squid nor octopus. About 8 inches long, this deep-water denizen was first described in 1903 by Carl Chun, a German teuthologist who identified it as an octopus, because it had—he thought—eight arms. Then another pair of thin arms was discovered, tucked into pockets outside the web that connects the eight arms. Taxonomically speaking, it hovers between octopus and squid in its own order, the Vampyromorpha.

Until very recently, it was thought that the vampire squid was a weak swimmer, and its weak-muscled, gelatinous body suggested drifting rather than darting as its primary method of locomotion. It has a highly developed statocyst, the organ that controls its balance, further enhancing the idea that *Vampyroteuthis* was an almost passive predator. Imagine the surprise of researcher Bruce Robison, watching the transmissions of a robot camera deployed in Monterey Canyon off the coast of California, when a specimen of *Vampyroteuthis* darted into view.

Discussing their observations in an article in the *New York Times* (Broad, 1994), Robison, Michael Vecchione, and James Stein Hunt were absolutely amazed. "The images completely blew my mind," said Vecchione. "Nobody suspected it acted like that—buzzing around in circles and swimming rapidly. Usually they've been pictured as drifting with their arms spread out." The article continued:

> The brownish-red creature swam with great dexterity, flapping its large thin fins that looked like wings. It trailed a long thin filament, whose function is unknown but probably sensory. A close-up view showed the animal perfectly still, slowly opening and closing its large arms with a sinuous rhythm, revealing them to be joined by thick webbing. Visible at the animal's front was a huge eye, eerie and blue. At the center of its web was a beady mouth, held motionless.

In recent years, scientists have discovered that squids—usually the California market squid, *Loligo opalescens,* the Eastern squid, *Illex illecebrosus,* or the most accessible of all, *Loligo pealei* and *L. plei*—make excellent laboratory subjects, and therefore we can learn a great deal about the biology and biomechanics of these fascinating creatures by observing them in captivity. But even though the giant squid shares many characteristics with its smaller relatives, it is not simply a larger version of the market squid. It is a subject unto itself, still keeping its secrets hidden from prying eyes by its mysterious habits and impenetrable habitat. Even so, we cannot discuss the radula or the tentacular clubs of *Architeuthis* without establishing some sort of a teuthological framework; before we can understand how the giant squid works, we must have some background information on its smaller, more accessible relatives. (It is no more accurate to compare the information garnered from the study of an 8-inch *Loligo* to the 60-foot *Architeuthis* than it would be to equate a mouse with a hippopotamus on the grounds that both of them are mammals, but because we obviously cannot study *Architeuthis* in captivity, and since most of the specimens we have seen have been dead or dying, we must make these comparisons—even if potentially some of them are forced or, in some cases, totally inapplicable.)

Although many squid species have been examined, there are still many that are known from only a few examples, and we have no idea how many remain to be discovered. (Roper, Young, and Voss identified twenty-five distinct *families* of the order Teuthoidea.) In his 1966 "Review of the Systematics and Ecology of Oceanic Squids," Malcolm R. Clarke identified 181 species, and wrote, "The taxonomic tangle prevents any but the most limited analysis of the ecological data."

Since Clarke's study, improvements in the classification of squids have been slow indeed. Progress was so slow, in fact, that by 1977, when Gilbert Voss's work was published in *The Biology of Cephalopods,* the author identified "only nine critical revisions of cephalopod genera or families" that had taken place since 1966. He lamented that "from the brief review of the present number of families, genera, and species, it can be seen that the systematics of the class is still far from the stage when the broad evolutionary picture can be seen."

In 1980, Malcolm Clarke published an exhaustive guide to cephalopods based on his analysis of the squid beaks found in the stomachs of sperm whales captured in the Southern Hemisphere. This study contained precise, measured illustrations of the beaks sampled, and using this criterion, he was able to take a large step toward identifying individual species.

In 1981, at a cephalopod workshop held in Melbourne, Australia, Clyde Roper picked up where his mentor, Gilbert Voss, had left off, and overviewed the systematics of cephalopods again. He listed the problems of the cephalopod systematist, many of which involve the complexities of preserving cephalopods for study: when fixed in formalin, octopuses contract so vigorously that they cannot be examined again; soft-bodied cephalopods become permanently molded in the position they are originally fixed in; and probably the most pressing problem, there was a scarcity of adequate specimens. "Cephalopods," wrote Roper, "frequently are very difficult to catch; because they are very perceptive, and very fast swimmers, they are able to avoid the nets. Capturing adequate samples is so difficult, in fact, that those of us who sample oceanic and midwater groups insist that we catch only the slow, the sick and the stupid."

In recent years, several comprehensive studies have been published, including the 1984 *FAO Species Catalogue of Cephalopods of the World* (Roper, Sweeney, and Nauen), and Kir Nesis's *Cephalopods of the World* (1982), both of which go a long way to resolving the general taxonomic disorder, but there are still many unanswered questions, many species still undescribed, and probably more undiscovered. One of the major problems in the taxonomy of squids is that they often change dramatically from the early to the adult stages. As Voss wrote, "These animals, with only a few exceptions . . . go through developmental stages that are so varied that many of the larvae and juveniles of both octopods and teuthoids defy identification with known adults. . . . In other words, attempts to identify a larva using keys and descriptions based upon adult specimens will invariably lead only to confusion." In recent years, many authors have synonymized various species, added new ones, renamed some, and eliminated others on the basis of similarities or differences, but the actual number of squid species remains an unknown. At the moment, the best guesses range between six hundred and seven hundred species.

All squids are decapods (as contrasted with octopuses, which are octopods), but the arrangement and proportion of the arms differ from species to species.* While most squids have eight shorter arms and two longer tentacles, others (such as *Chiroteuthis*) have six arms of varying length, two heavier, longer arms, and two tentacles that are as much as ten times as long as the body. The majority of species have developed tentacles that are long, slender appendages with expanded, club-like ends, but in some cases the tentacles are the same length as the arms, differentiated only by the flattened *manus* at the end. In some species there are lateral expansions, or "swimming keels," on the outer surface of the third pair of arms, while in others, the arms are joined by a web. And there are some species that start out with ten appendages, but because the tentacles disappear as the animal matures, we have the oxymoronic eight-armed decapod. Although the pattern varies, species are identifiable by the arrangement of the suckers on the arms and the tentacles, some of which have retractable claws, and most of which are equipped with toothed rings.

Squids catch their prey with their tentacles, which can be partially retracted when not in use. They draw the victim in toward the center of the circle of arms,

---

*Because the arms can be different for each species, and even in the same species, a numbering system has been established that enables teuthologists to identify specifically which arm they are referring to. As the squid faces you (with the funnel on the ventral surface), the uppermost pair of arms (on a plane with the "top of the head") is numbered I, the next pair downward is II, then III, and IV refers to the arms nearest the funnel. The two tentacles are always counted separately, and are located between arms III and IV.

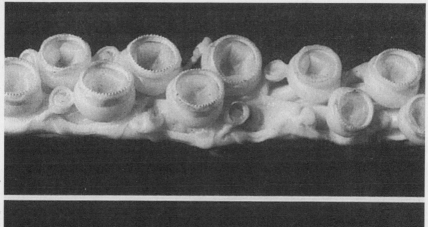

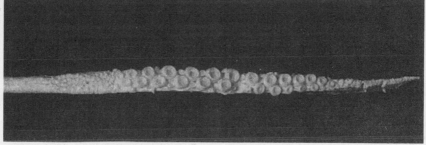

Detail of suckers from tentacle club of the 1995 South Australian Architeuthis.

where the powerful, parrotlike beak is located. The eight arms are thicker than the tentacles, and are equipped with a double row of suckers that decrease in size toward the tips. (Because of the location of the mouth, the inner sucker-bearing surface of the arm is referred to as the *oral* surface, and the outer side the *aboral*.) The tentacles are always positioned between the third and fourth arms, known as the ventrolateral and ventral arms. (The other two pairs are the dorsal and dorsolateral.) In cross section, the tentacles are round, while the arms are roughly triangular, with the suckers on the flattened oral surface. In many teuthids, the tentacles are equipped with a "locking apparatus," consisting of tubercles and suckers that fit together like snaps, allowing the tentacles to be held tightly together when extended. The tentacles are longer and much smaller in circumference at the base than the arms, and function like a pair of tongs or pliers, grabbing the prey and pulling it into the crown of arms, where it is then transferred to the mouth. Unlike those of the octopods, the suckers of squids are on stalks, and can be moved independently. In many species, the suckers on the arms and the tentacles are equipped with a ring of chitinous "teeth," which help the squid keep a grip on its often slippery prey.

Some suckers have teeth, and some do not. But otherwise, are they more or less the same? Andrew Smith (1996) does not think so. In the laboratory, Smith tested the pressure created by octopus and squid suckers and concluded that "the strongest suckers belong to the fast-swimming, open water species in the decapod suborder Oegopsida." In fact, he found that squid suckers created such tremendous pressure that the sucker would be torn off the stalk before it could be pulled free of the surface it had adhered to. He also found that "smaller suckers . . . would produce even greater pressure differentials [but] the reason for the greater strength of small suckers is unknown."

While it does not conform to our customary definition, usually having something to do with a protuberance at one end of the body, a squid has a head. It is separated from the body by a neck (technically known as the nuchal constriction), and it is usually smaller in diameter than the body. (In the Histioteuthidae, however, the opposite is sometimes true.) Many larval squids can retract their heads partway into the mantle; the only adult animals that retain this ability are species of *Onychoteuthis*. Most of the width of the teuthid head is a function of the size of the eyes, which, in some species, are enormous. (In some of the cranchiids like *Taonius* or *Galiteuthis*, the head seems to be almost all eyes.) The main support structure inside the head is the cartilaginous head capsule, which houses the eyes, the brain, and the statocysts, and through which the esophagus passes. This narrow passage means that squids cannot swallow large pieces of food, and have to chop the food up with the beak or grind it down with the radular teeth. Chemical recognition of prey (the sense of smell) has not been extensively studied, probably because squids are such visual animals. They do have supposed olfactory organs, which are positioned on the sides of the head near the neck. At the posterior end of the head is the body (mantle), but at the other end is a circle of arms and (usually) a pair of tentacles, surrounding the squid's eating apparatus.

The mouth of a squid is referred to as the buccal cavity. On a muscular tonguelike projection known as the odontophore is the radula, a chitinous ribbon whose surface is covered with transverse rows of backward-pointing

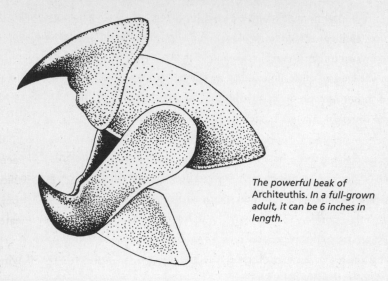

The powerful beak of Architeuthis. *In a full-grown adult, it can be 6 inches in length.*

teeth, which help move the food through the gullet and grind it down.* The radular teeth are usually arranged in rows, and they can be used to differentiate species.

The beak of the squid is located at the center of the corona of arms, and consists of an upper and lower mandible, set into a muscular configuration known as the buccal bulb. Using the buccal musculature, the squid can rotate and protrude its beak. The beak is hard, like the beak of a parrot, but unlike the bird's beak, the upper mandible is the smaller of the two and fits inside the lower. In young cephalopods, the beak is a semitransparent amber color, but as the animal matures, the beak darkens, becoming dark brown or black in sexually mature adults.

In examinations of the stomach contents of sperm whales, when the soft parts have been digested, the beak is the most important diagnostic tool. In "Cephalopoda in the Diet of Sperm Whales of the Southern Hemisphere," Malcolm Clarke wrote,

---

*When A. E. Verrill was describing a specimen of *Architeuthis monachus* for the 1875 issue of the *American Journal of Science,* he made what he later described as a "serious mistake in respect to the lingual ribbon, or odontophore, of the specimen." He mistook the radula for the odontophore because in this specimen, the former occupied the position normally reserved for the latter. He was able to rectify the mistake in a later edition of the journal, and he wrote, "I was fortunate to find, several months after my papers were printed, the genuine *odontophore* among the dirt and debris that had remained in the bottom of the can in which the specimen had originally been sent from Newfoundland."

While relatively few complete cephalopods are collected from stomachs, the hard chitinous mandibles or "beaks" are largely unaffected by digestion and accumulate in the stomach. Beak samples may be very large (up to 18,115 beaks in one Durban whale) and, if they are identified, can provide different and better information on the whale's diet and on cephalopods than can the relatively few complete specimens (rarely more than 30 in a stomach).

Clarke also wrote a paper entitled "New Techniques for the Study of Sperm Whale Migration," in which he described how the identification of specific squid beaks in the stomachs of captured sperm whales would enable researchers to identify the migration routes of the whales. If, for example, the whales consumed certain Antarctic squid species, such as *Mesonychoteuthis* or *Moroteuthis knipovitchi,* and the whales were caught outside this range—say, off South Africa—it would show where the whales had been.

Sperm whales eat squids, but what do squids eat? Depending on their size and feeding equipment, squids eat everything from planktonic animals to crustaceans, fishes, and other squids. Not surprisingly, the stronger species feed on larger prey. In his 1970 discussion of *Dosidicus,* Nesis wrote that the jumbo squid feeds mainly on lantern fishes and other squids. It is, he reported, "a schooling nektonic predator, which eats any prey that moves, provided only that it is abundant and of convenient size. . . . They even devour garbage from a ship, cucumber rinds, banana peels, and the like." Hanlon and Messenger describe all cephalopods as "voracious carnivores, feeding by day or by night on a wide variety of live prey that they detect mainly with their eyes or by touch."

We have only the vaguest idea of the prey of *Architeuthis.* Because the radula shreds the food, few identifiable pieces have ever been found in the stomachs of stranded specimens. Furthermore, most of these specimens have been so battered by surf or rocks that the stomach contents were not available for examination. As an indication of how little is known about the feeding habits of *Architeuthis,* in a paper published in the *Canadian Journal of Zoology* in 1969, John Pippy and Frederick Aldrich, having identified a parasite in the muscle tissue of a giant squid that washed ashore in Newfoundland, and noting that this parasite *(Hepatoxylon trichiuri)* is known only from bony fishes and sharks, wrote, "Unfortunately, there

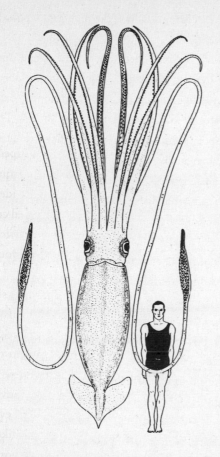

In an article called "Marauders of the Sea" in a 1935 issue of National Geographic, the artist added a man to a copy of A. E. Verrill's drawing of the giant squid that washed ashore at Trinity Bay, Newfoundland, in 1877. The credit line for the drawing reads: "Drawn by Zhneya Gay after A. E. Verrill Illustration."

is no information on the feeding habits of the giant squids. . . . This report is the first indication of what may be the food of the architeuthids, namely species of teleosts or selachians." In other words, giant squid might eat bony fishes or sharks—not a particularly helpful contribution. Only in 1995 was someone able to examine the stomach contents of a giant squid. Three males had been trawled up off southwestern Ireland, and their stomachs contained the remains of various fishes, prawns, octopuses, and other squids (Lordan, personal communication, 1997).

While the squid has a "top" and a "bottom" (*dorsal* and *ventral* in anatomical parlance), it does not really have a front and a back. Yes, there is a tail at one end, and a cluster of arms at the other, but the squid can move through the water with either one as the lead element. (The tail end is usually referred to as the posterior, and the tentacle end as the anterior.) Squids propel themselves—in whatever direction they choose—by a form of "jet propulsion," by shooting water from the funnel, a short, hoselike organ that projects from the mantle below the head. The squid can rotate the funnel in any direction, and by ejecting jets of water from it, move in the direction of its arm tips or, just as easily, in the direction of its tail fins. (As seen in captive, smaller squids, when the animal moves in the direction of its arms, it clusters them together in order to reduce the drag that would be created by a flailing bunch of appendages.)

To ensure that the ejection of water is controlled, the mantle is equipped with a ridge or knob (the "mantle-locking cartilage") that fits into a corresponding pocket or groove (the "funnel-locking cartilage") in the funnel component, to seal the mantle when water is to be ejected from the funnel. This arrangement differs from species to species, and can be used as a diagnostic tool to differentiate the species. The locking cartilages of *Architeuthis* are described as "straight and simple." There is no "forward" or "backward" for the giant squid, but obviously disdaining such terrestrial distinctions, it can move powerfully through the water in any direction it chooses. (Just *how* powerfully is another unanswered question.)

Most species of squid have a pair of fins at the tail end of the mantle, which are used for locomotion by an up-and-down undulation. In some genera, such as *Dosidicus* and *Ommastrephes,* these fins are large and powerful, indicating that the squid uses them for propulsion. *Taningia,* a powerful swimmer that can reach a length of 7 feet, sports a pair of wide, powerful fins that make it look not unlike a ray with arms. In many other species—*Architeuthis,* for example—the fins are relatively small, suggesting a reliance on other forms of power. In 1931, Paul Bartsch wrote: "Inch for inch, the squids will compete in swimming power with any other creature that lives in the sea." More recently, at Dalhousie University in Halifax, Nova Scotia, Ron O'Dor has been studying the smaller squids, and has come up with some remarkable observations. O'Dor refers to the squids as "invertebrate athletes" and "the elite cadre of the most athletic class of molluscs." In a 1989 study (with R. E. Shadwick) called "Squid, the Olympian Cephalopods," O'Dor wrote that "the inherent inefficiency of jet propulsion . . . forced squid to develop their remarkable athletic prowess. Squid not only migrate actively over thousands of kilometers, their power outputs and oxygen consumptions are higher than anything else in the sea." Squids are drag racers compared to fishes, which cannot accelerate as fast, but can sustain a pace for a much longer time. Jet propulsion also moves more water over the gills, enabling the squids to extract oxygen from the water more effectively than fishes. To transport oxygen, squid blood uses the copper-containing pigment hemocyanin rather than the hemoglobin used by vertebrates, but it is less viscous, and therefore easier to pump. (Hemocyanin means "blue blood," and oxygenated squid blood is almost colorless, with a bluish cast.)

In squids, as in all cephalopods, the circulatory system is enhanced by the presence of two branchial hearts that contract rhythmically to force blood into the two gills, which are attached to the inner wall of the mantle by a thin membrane. Breathing consists of "inhaling" water into the mantle cavity, where the gills extract the oxygen, and then passing the water out through the funnel. The oxygen content of the water in which they live is very important to cephalopods; if it is low, they quickly lose their strength, become flaccid, and die.

Throughout the course of their evolution, squids have developed various adaptations to enable them to compete with fishes. "There are too many things that fish do well that squid either do poorly or cannot do at all," wrote O'Dor and Webber in 1989. "On the other hand, there appear to be some things that squid can do better." The squids devote a higher portion of their body weight to the nervous system than cold-blooded vertebrates such as fishes (approaching some reptiles in that regard), and the neural control of the circulatory system probably accounts for the incredible athletic feats they can perform. The large and complex brains of some cephalopods enable them to implement a process that fishes lack entirely: the direct nervous control of color and pattern. Squids use color for defense, and there is even an indication that they may have evolved a pattern-based color "language" that affects their social organization. ("One certainly has the impression," wrote O'Dor and Webber in 1986, "that there is more going on than a simple random association of equal individuals as is found in most squid schools.")

As Frank Lane has written, "Cephalopods are not generally considered to be colorful animals. It is therefore all the more surprising to find that they surpass even the famed chameleon in the speed and variety of their color changes." The skin of most species is equipped with a dense field of round, elastic-walled cells known as chromatophores; these can be expanded or contracted all at once or in sequence, producing a dazzling variety of spots, stripes, or, most spectacularly, a moving wash or ripple of changing hues. In various species, the individual pigment cells come in many colors, including red, orange, yellow, brown, and black, but *Architeuthis* probably does not possess such variety.

Smaller squids employ their ability to change color in a variety of ways. Their primary defense mechanism is flight, but they can also camouflage themselves to blend in with their environment. They can be light-colored in shallow water or in bright sunlight, darker in deep water or at night. If this defense does not work, an individual will flee, but this flight may also be accompanied by a lightning-fast color shift. Another defense mechanism is "deimatic behavior," which involves the rapid change of what Hanlon and Messenger prefer to call "body patterns" (as opposed to "color change"), in order to bluff, frighten, or startle the predator. If the squid can interrupt the normal attack sequence of a predator—even for a millisecond—it may be able to gain enough time to make its escape. In this process, a squid might change color several times, adopt a pattern of prominent eyespots or stripes, or even alter its texture by erecting bumps or nodules on its skin.

As a deep- to midwater inhabitant (we think), *Architeuthis* would have little occasion to match the bottom, and it is almost impossible to imagine a 60-foot-long animal trying to hide in the open ocean. (Of course, if the squid spends most of its time at depths where no light penetrates, the problem of "hiding" may be academic.) Although chromatophores have been found in the skin of the giant squid, they are not so numerous or varied as those of their smaller cousins. Their "palette" is probably a comparatively narrow one, ranging from pale gray to dark purple, and their color-change ability may be primarily employed in mating rituals.

Since squids have no mechanisms for making sounds, it is assumed that communication among individuals and groups is conducted—at least in part—by color- or pattern-changing. In competition for the attention of a female, males will run through a repertoire of color changes that are exclusive to males, "dueling" with color for the hand(s) of a mate. The female also uses body patterns to signal her choice, or perhaps to induce more flashing by the competing males.

Many species of squid are also equipped with light-producing organs known as photophores. These are, as Malcolm Clarke put it, "special photogenic cells, often equipped with a reflector that can point the light in one direction, redirect it to an area remote from the source, or spread it out over a surface. Sometimes the light passes through filters which change the wavelength of the emitted light and sometimes it escapes through half-silvered tubes to increase its angle of emission. These devices can impart a variety of effects, from an overall dull glow or a bright wooly, ethereal effect, to sharp pinpoints of light or a torch-like beam." Among the primary uses of the photophores is counterillumination. To prevent a predator from picking out the squid silhouetted against a sunlit surface, it can illuminate its underside, effectively "downlighting" itself and eliminating its shadow. (Some species can also emit a cloud of luminescent bacteria as a flashing pseudomorph, designed to seriously confuse a potential predator.)

Aboard the *Valdivia* expedition in 1899, the German teuthologist Carl Chun (who published some of the most important and comprehensive studies of squids ever written, including the first description of *Vampyroteuthis*) was observing some individuals of the small squid *Lycoteuthis diadema* in a container on deck:

> Among the marvels of coloration which the animals of the deep sea exhibited to us nothing can be even distantly compared with the hues of these organs. One would think that the body was adorned with a diadem of brilliant gems. The middle organs of the eyes shone with ultramarine blue, the lateral ones with a pearly sheen. Those towards the front of the lower surface of the body gave off a ruby-red light, while those behind were snow-white or pearly, except the median one, which was sky-blue. It was indeed a glorious spectacle.

But, as with so many aspects of teuthid biology, luminescence is still poorly understood. In a 1977 essay on this subject, Peter J. Herring of the Institute of Oceanographic Services in Surrey, England, wrote:

> So much of the information on luminescence in cephalopods and fish is still purely anatomical, and so bewildering in its variety, that functional interpretations almost certainly err on the side of conservatism, an inevitable consequence of our restricted concepts of our understanding of the normal lives of these remarkable animals. Any purpose that is fulfilled by colour or pattern in the illuminated terrestrial or coastal environment can also be achieved by luminescence in the dark of the deep sea. The cephalopod inhabitants of this environment almost certainly make far more extensive and varied use of their impressive luminescent abilities than we can presently envisage.

In the early history of submersible exploration, various observers had spotted cephalopods through the portholes. William Beebe, describing a dive in the bathysphere in 1930, wrote that "a school of large squids . . . their great eyes, each illuminated with a circle of colored lights, stared in at me—those unbelievably intelligent yet reasonless eyes backed by no brain and set in a snail."* Diving in

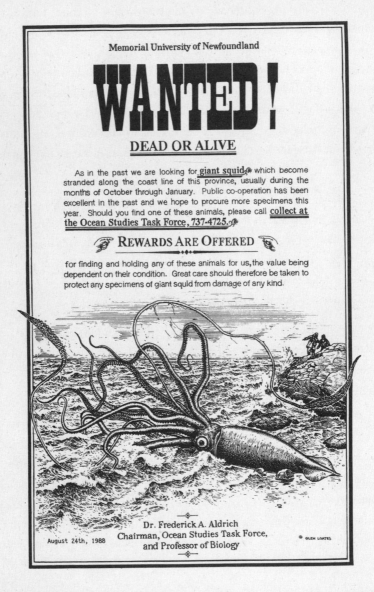

Issued by Frederick Aldrich and drawn by Glen Loates, this poster shows a giant squid flailing on the Newfoundland shore.

the French submersible *FNRS-3* in the Mediterranean in 1954, Jacques Cousteau wrote that he saw "a beautiful squid, which stops for a second, as if dazzled by the searchlight beam. . . . I see clearly its rocketlike head and its 10 arms. It is about 1½ feet long, and it leaves a blob of ink. The ink is white."

Even though they can change color with remarkable speed, the "base color" for most squids is a somewhat translucent silvery gray. (In addition to the photophores and chromatophores, some squids have iridophores, or reflecting cells, that are responsible for the iridescent blues and greens sometimes found around the eyes or on the dorsal mantle surface.) In some species that have been extensively studied in captivity—*Loligo pealei,* for example—the interior organs are faintly visible through the skin unless the animal decides to affect a different color scheme.

Generally battered and torn by rolling on the beach, the skin of many of the giant squid that have been examined often showed evidence of red or brown chromatophores, and this has led to a profusion of drawings—usually depicting the squid in a battle with its "mortal enemy," the sperm whale—where the squid is blood-red. A nice conceit, but probably incorrect. In *Octopus and Squid,* Cousteau and his crew observed that common octopuses were usually mottled brown in color, but turned redder as they got "angrier." Moreover, the Pacific giant octopus *(Octopus dofleini) is* dark red, so it was not that much of a leap to assume that the giant squid might be the same color. In *Beast,* Peter Benchley introduces the eponymous protagonist as "purplish maroon," and later on, as the animal becomes more agitated, it turns "a lighter, brighter red—not a blood red, for so permeated was its blood with haemocyanin that it was in fact green, but a red designed by nature purely for intimidation."†

---

*Beebe was dead wrong about the brainlessness of the squid, since even at that time the teuthids were known to exhibit remarkable cognitive abilities. In 1917, Paul Bartsch, curator of marine invertebrates at the Smithsonian, wrote, "The largest, the most highly organized, as well as intelligent, and therefore, most interesting invertebrate creatures of the sea belong to the class of organisms known as cephalopods." As a zoologist, Beebe should certainly have known that while a squid is indeed a mollusk, it is certainly not a *snail.*

†Oxygenated squid blood is pale blue, and does not affect the coloration of the animal. Even if nature designed red "purely for intimidation," the squid's prey would not be intimidated, for at the depths where the giant squid hunts, all colors are removed from the spectrum (red, with the shortest wavelength, is the first to go), and with exception of bioluminescence, everything appears black.

While almost all squids can change color, very few of them maintain a constant dark coloration, and it is more likely that *Architeuthis* improves its chances for remaining undetected while awaiting its prey by assuming a neutral coloration, to accompany its neutral buoyancy. The cryptic coloration, the absence of lights, and the ability to remain motionless in the water (most fishes are extremely sensitive to movement around them through their lateral-line system) suggest a picture of *Architeuthis* hovering menacingly somewhere in the water column, head and arms angled downward. This picture does not jibe with the conventional image of *Architeuthis* zooming around looking for sperm whales to wrestle or people to eat (see Benchley's *Beast*, for example), but it is frightful enough to envision the giant squid lurking almost invisibly in the blackness of the depths, its amazing eyes able to pick out the tiniest light or movement, its muscular tentacles shooting out to capture its unsuspecting prey. We do not have to embellish *Architeuthis* with more frightening attributes; it does quite well with those it already has.

Although the cephalopods are generally considered among the more primitive animals (they are classified with the mollusks, along with snails, clams, and oysters), Paul Bartsch wrote that "the mollusca rank at the top of all invertebrate life in complexity of organization and intelligence, as they certainly do in size, ferocity, and speed of movement." Clams and oysters do not have prominent eyes, but the eyes of the cephalopods are among their most remarkable features. The biological principle of convergence, where similar characteristics have evolved in totally unrelated animals, is marvelously demonstrated by comparing the eye of the squid and that of the higher vertebrates. Both have eyes in which there is a large posterior chamber filled with aqueous humor, which contains a pupil that can expand or contract; both have an iris diaphragm and an adjustable lens set in a muscular ring. Both types of eyes have dark pigments that act as light screens, and both have fibrous coats to retain the eye's shape. Where vertebrate eyes have rods and cones, however, the eye of the squid has long, thin receptor cells that bear microvilli, which contain the visual pigment. Half the receptor cells have the microvilli oriented vertically, and the other half horizontally, which indicates a sensitivity to polarized light. They see well, but there is evidence that they do not see color.

Within the order Teuthoidea, there are two suborders, differentiated by the presence or absence of a transparent corneal membrane over the eye. The

Myopsida have such a membrane; the Oegopsida—the group that includes *Architeuthis*—do not. Of the twenty-five known families of squid, twenty-three are oegopsids. Except for those species with eyes on stalks, squids' eyes are not binocular; each one sees what is on its side of the head. "Whereas the human eye has a point of focus on the retina," wrote Harry Thurston, "the squid [eye] has an equatorial band where it is able to focus, thus making the squid's vision, theoretically, twice as good as a human's. Sight, more than any other sense, dominates a squid's life."

The most remarkable thing about the eyes of *Architeuthis* is their size, but some deep-water species have their eyes on stalks; some have eyes with built-in lights; and some have a right eye that is completely different from the left. Squids of the genus *Histioteuthis* have a relatively gigantic left eye and a right one that is approximately a quarter as large. In addition, the smaller eye has a series of photophores—as many as eighteen in some species—around the periphery, the function of which is also mysterious. The subject of these dramatically different eyes has long intrigued teuthologists. In 1968, E. J. Denton and F. J. Warren suggested that the larger eye is specially adapted for looking downward, because it can better pick out small spots of light. Richard Young (1975) collected several specimens of *Histioteuthis dofleini* off the Hawaiian island of Oahu and put them in a tank to observe them. When he examined the eyes, he concluded that the larger eye (the left one) is a *tubular* eye that is usually directed upward, while the smaller right eye faces downward. He wrote that "the larger upward-looking eye detects down-welling surface light, the small eye does not; rather, it detects bioluminescent light. . . . The compact arrangement of photophores around the smaller eye, combined with the modifications of the retina, suggests that countershading is not the only function of these organs. These ocular photophores are ideally located to produce a strong beam of light that would illuminate the portion of the environment that is surveyed by the thicker portion of the retina." *In other words, these photophores may function as a searchlight.* (Young's conclusion, my italics.) A squid with a searchlight!

We do not know where *Architeuthis* hunts, but we do know that it has the largest eyes in the animal kingdom. Akimushkin is on record as declaring that the eye can reach 400 mm (15.6 inches) in diameter. (An ordinary dinner plate is 10 inches in diameter.) Since the eyes of stranded or digested squids are among the first elements to deteriorate, hardly anything is known about the actual structure of the eye of the giant squid.

A 1988 article on squids by Malcolm Clarke opens with these words: "To many biologists the squid conjures up a vision of an unusually large nerve fibre with graphs issuing from one end while electrodes are applied to the other." The "unusually large nerve fibre" is another wonder of squid biology, for it can be one-tenth of an inch in diameter, as compared with the largest human axon, which is only one one-thousandth of an inch. The size of these giant axons enables the squid to transmit messages to its muscles substantially faster than any other creature; the squid's ability to respond to a particular stimulus can be almost instantaneous. For human neurological research, these giant axons are much easier to study than those of most other animals.

The cephalopod brain consists of a concentration of nerve ganglia, and it is enclosed in a cartilaginous "skull." Although only a few species of squid have been examined in detail, many have been observed to have sensory papillae behind the eyes, which may function as organs of taste or smell. Some even have sensory cells all over their bodies, which are thought to be sensitive to changes in temperature. Octopuses survive far better in captivity than squids, so much more work has been done in testing their capabilities. But if we could figure out what questions to ask and how to ask them, we might find the squid the intellectual equal of its eight-armed relative.

Martin Moynihan has made a specialty of studying the behavior of squids, particularly the Caribbean reef squid, *Sepioteuthis sepioidea*. Snorkeling in the waters of the San Blas Islands, off the east coast of Panama, Moynihan and A. F. Rodaniche observed the little squid (with a maximum mantle length of 8 inches) in every possible activity, from courtship to predation, and have concluded that they are "intelligent" enough to make conscious choices regularly. In their 1982 *Behavior and Natural History of the Caribbean Reef Squid*, they wrote:

> Reef squids and other cephalopods are forced to make choices constantly. The elegant experimental studies of the brains of *Octopus vulgaris* [common octopus] and *Sepia officinalis* [common cuttlefish] have focused upon decision-making, usually the decision between attacking and not attacking a possible

prey, a crab or a shrimp, under controlled conditions in the laboratory. Doubtless the experimental animals take the problem seriously, especially when mistakes are negatively reinforced by electric shocks. Yet for many cephalopods, including *S. sepioidea*, selection of prey is neither the most frequent nor the most difficult of the problems with which they attempt to cope under natural conditions in the wild. They are more often concerned with social or sexual companions or possible predators. The decisions they face all the time are literally matters of appearance. Is it better to be Dark or Pale? Or Barred or Streaked? Or Yellow or Blue? Or upside-down or downside-up? Among cephalopods, as perhaps other animals, these are the questions that decide the future of individuals and of species.

Hanlon and Messenger, in their *Cephalopod Behaviour*, wrote, "We have seen that cephalopods are 'brainy' animals and their behaviour is complex and diverse. They not only have good senses with which to explore their world and select the best course of action for survival, but they have memory stores that can provide up-to-date information on which to base their decisions. In such short-lived animals, one might have predicted more rigid, pre-programmed behaviours, rather than the flexible responses based on memory."

When it comes to *Architeuthis*, it is important to realize that while it may be larger than the others, it still is a squid, and while the details may differ from species to species, the general structure and functions are the same—the definition of an order. They are all built along the same general plan, and all have variations of the same equipment. We don't know if *Architeuthis* thinks, or if it does, what it thinks about. We do know that as the largest squid species, it has the largest brain, and for the most part, a larger brain equals more complex responses. (The sperm whale has the largest brain of any animal that has ever lived.) That the giant squid has remained unseen underwater for so many centuries might be nothing more than coincidence—divers and submersibles in the wrong places at the wrong times—but it could also be that the giant squid doesn't *want* to be seen. Admittedly, this is wild speculation, but an animal whose major predator (the sperm whale) is a large, more or less cylindrical object might be inclined to stay as far away as possible from strange, more or less cylindrical objects, for example, submersibles and robotic camera housings.

Although we rarely think about it, we (and all other vertebrates) have developed a sophisticated system of orientation to keep us aware of our position in three-dimensional space. As the University of Texas zoologist Bernd-Ulrich Budelmann put it in 1980, "Whereas other stimuli, such as odor, taste, light, and sound may change or even be absent during periods of time, the gravitational field has a unique feature: during the life span of an organism, it is constant in magnitude and direction." In mammals the equilibrium system consists of the vestibular apparatus of the inner ear, but cephalopods, having no ears, inner or outer, have had to devlop an alternative system. In the octopus, cuttlefish, and squid, the equilibrium receptor system consists of fluid-filled vesicles called statocysts, which have calcareous particles called statoliths suspended within them. The paired statocysts are embedded in the cartilaginous brain capsule, and orient the animal with regard to gravity and movement, producing the mysterious sense we call balance.

In many areas, cephalopods have developed characteristics to enable them to compete with the bony fishes; advanced locomotion (that is, jet propulsion), efficient energy consumption, and the development of superior vision are among these. Their ability to change color instantly is a defensive adaptation that may be more effective than flight, and is certainly superior to the slow camouflage technique of fishes such as flounders. In some species of cephalopods, the statocyst system closely resembles the inner ear and lateral-line system of the bony fishes, which enable them to maintain their balance and sense movement in the water,* and the statoliths of the octopus and squid correspond to the otoliths of fishes.

Although we have come a long way from what Jacques Cousteau called the "silent world" in 1953, we still do not automatically associate the undersea environment with noise. (For cetaceans, particularly the odontocetes, sound is the primary means of navigation, communication, prey location, and even prey debilitation. See pages 157–162 for a discussion of the sperm whale's reliance on

---

*It has recently been shown that some cephalopods also have a "lateral-line" system. Examining parallel lines of epidermal cells in embryonic cuttlefish, Budelmann, Riese, and Bleckmann wrote, "We have now demonstrated that these lines of cells serve to detect small water movements and thus are a further example of convergent evolution between a sophisticated cephalopod and a vertebrate sensory system."

sound.) There are fishes that make noises, shrimps that snap their claws, and even some squids that have been recorded to make crunching noises while eating. But do the cephalopods, with no known structures for hearing, respond to sound? And if so, with what?

In 1985 Martin Moynihan wrote an article for the *American Naturalist* entitled "Why Are Cephalopods Deaf?," in which he suggested that deafness "might be an adaptation to cope with, or adjust to, odontocete attacks." Because cephalopods are susceptible to the "bombing" of toothed whales, deafness might have been an adaptation to avoid the whales' predation. He also wrote that cephalopods probably don't need hearing as a protective device because their vision is so good. However, Moynihan knew that his theory might be weak, and it was not long before Roger Hanlon and Bernd-Ulrich Budelmann took issue with it.

In a 1987 issue of the same journal, they wrote "Why Cephalopods Are Probably Not 'Deaf,'" in which they said, "Surprisingly, Moynihan did not consider the available morphological and physiological data." Among these data: blind octopuses have been observed to respond to vibrations caused by tapping on a tank, and in one set of experiments, when a hidden experimenter knocked on a tank containing small squids, they instantly changed color and jetted backward. Hatchling squids, examined under scanning electron microscopes, were observed to have ciliated cells on the head and mantle, and a 1963 study of the common octopus by Maturana and Sperling showed that the statocysts contain hair cells that are sensitive to low-frequency vibrations.

To justify the existence of an auditory capability, Hanlon and Budelmann postulated the ecological necessity for such a system:

> Much predation by cephalopods occurs at night or in mesopelagic or bathy-pelagic zones, where light is greatly limited or lacking. Many cephalopods also live in murky water. Vision, therefore, is a restricted resource that may not allow detection of a fast-moving predator until it is too late. Hearing is a far better distance receptor, and it does not depend greatly on the time of day, depth (i.e., light intensity), or water clarity. It seems likely from an evolutionary standpoint that, rather than being deaf, cephalopods have developed a sensory system for underwater wave perception that is . . . capable of detecting the approach of prey

items or predators at such distances that appropriate behavioral responses can be implemented.

According to Hanlon and Budelmann, "hearing is nearly impossible to define for aquatic animals, because sound and vibrational stimuli are essentially equivalent in an aquatic medium," so until more detailed information on the hearing capabilities of octopuses and squids appears, we can assume that they are not deaf. In 1990, Packard, Karlsen, and Sand wrote that "cephalopods seem to be well equipped for detection of water-borne vibrations. Ciliated sensory cells that look like mechanoreceptors have been found in epidermal lines on the head and arms of squids and cuttlefish, and Budelmann and Bleckmann (1988) have recently demonstrated that local water movements evoke microphonic potentials in the head lines of these animals." They also wrote, "In the infrasound [low-frequency] range, cephalopods can hear quite well."*

The question of the squid's response to the whales' sound bursts is a complex one. One of the more persuasive arguments against squid "hearing" was put forward by Michael Taylor, in an essay in *Nature* in 1986. (It was a comment on Moynihan's 1985 article on why cephalopods are deaf.) Taylor observed that since the whale has to locate the squid by sending out bursts of sound energy and then waiting for the echoes to tell it where the prey actually is, a squid that can hear would hear these echolocating sounds even before the echoes returned to their sender, and be warned well in advance of an impending attack. However, many would argue that even if the squid *could* hear the broadcast sounds of the whale, it probably could not ascertain the direction from which the sounds were coming, and could not, therefore, take the necessary evasive action. There is no question about the acuteness and directionality of odontocete hearing (mostly

---

*How about squids *making* sounds? In a talk given in 1857, our old friend Professor Steenstrup had this to say on that subject: "There are not a few observations to the effect that cuttle-fishes have really uttered very loud sounds; in this respect he mentioned Barbut . . . who compares their noise to the grunting of a hog; Gundløg Svensen . . . who says of *Ommatostrephes* that it 'squeaks' and 'moans miserably whether it swims in the water or lies on the beach,' to say nothing of the remarkable statements by Olavius, who points out [that] the mollusk *Sepia loligo*, whose miserable cries can be distinctly heard half a mile away when it is chased ashore by the coal-fish *(Gadus virens)* or by any other equally dangerous pursuers." In an 1881 paper, Steenstrup included *Gonatus* as a species capable of uttering a loud sound, "as maintained by the Eskimos from several parts of Greenland."

observed in captive bottlenose dolphins), so the whale still holds a significant auditory advantage.

Hearing—or lack of it—relates directly to our discussion of *Architeuthis*. Even though the giant squid is not the predominant prey item of the sperm whale, it stands to reason that a larger prey item will provide more nourishment than a smaller one, and it behooves the predator to structure its hunting effort to maximize its caloric intake. If it takes approximately the same amount of energy to capture a smaller squid as a larger one, it makes sense to assume that the whale would take a giant squid over a smaller one. It may have to do with light, or sound, or the movement of water, but in the depths of the world's oceans, as the giant whales dive in pursuit of the giant squid, a balance between predator and prey has to be achieved; in order to prevent extinction of the entire race, the squid have somehow to be aware of the presence of the whales, and some of them have to escape.

Despite the mystery attending the giant squid, its smaller relatives play an important role in the study of human physiology. The smaller squids, such as the

Battles between giant squid and sperm whales are believed to occur because the whale is trying to eat the squid, and not vice versa.

foot-long *Loligo,* have comparatively huge nerve fibers, which are extremely important in neurological studies. In 1909, L. W. Williams of Harvard University noted the size of these fibers, but he did not recognize their potential for human studies. He wrote, "The very size of the nerve processes had prevented their discovery, since it is well-nigh impossible to believe that such a large structure can be a nerve fiber." Compared to the largest human nerve fiber, which is one one-thousandth of an inch in diameter, the fiber of *Loligo* can be one-tenth of an inch, about the size of a small pencil lead, and that of *Architeuthis* may be even larger. For the squid, larger fibers translate into faster conduction of impulses, accounting for the meteoric reaction time of the smaller species, and also their rapid-fire color changes, but these axons are also critical to humans.

It is not clear as to whether giant squid have *giant* giant axons. In his 1977 essay "Brain, Behaviour and Evolution of Cephalopods," J. Z. Young (who spent his fifty-year career concentrating on the neurology of squids and octopuses), having dissected a specimen that washed ashore at Scarborough in 1933, wrote, "None of the nerves examined contained the exceptional large fibres reported by Aldrich and Brown (1967). We may conclude that *Architeuthis* is not an especially fast-moving animal. This would agree with evidence that it is neutrally buoyant with a high concentration of ammonium ions in the mantle and arms (Denton, 1974)."

Woods Hole biochemist Francis C. G. Hoskin has been working on a rather unusual aspect of squid neurology. He has discovered that the nerves of cephalopods contain an enzyme that can destroy and thus render harmless a group of compounds that the popular press refers to as "nerve gases," extremely toxic because they poison an enzyme that is essential to nerve function. (Nerve gases, which include sarin, the culprit in the 1996 Tokyo subway deaths, are actually liquids, but they do have vapor pressures, and are therefore commonly known as gases.) In 1965, Hoskin and his colleagues thought that since the squid's giant axon does not have the insulating layer of myelin with which our own—and most other—nerves are wrapped, they could use these axons to test the effects of certain poison gases on nerve functions. When they subjected the axons to nerve gases, they were astonished to observe that something in the axon totally neutralized the effects of the gas.

The answer to the obvious question of why a squid, whose ancestors developed perhaps half a billion years ago, should maintain a defense mechanism against a substance that was first manufactured in 1854 is not so simple, but it fits well within current evolutionary theory. Genetic modifications are not "designed" to resolve particular problems (birds and insects did not evolve wings because they needed to fly), but rather, existing or developing modifications were appropriated for eventually advantageous purposes. It is believed that feathers developed from modified scales that were used for thermoregulation, and were later employed for flying. The squid developed this unusual (and unexpected) immunity to nerve gas, but as Hoskin wrote, "Perhaps the nerve-gas-detoxifying enzyme is really there for some simple, well-known purpose that we have missed, but this seems unlikely in view of the specificity of enzyme reactions." In cephalopod nerves, half the chloride content has been replaced with an ion called isethionate, an alcohol molecule with sulfonic acid at one end. Hoskin speculated that the squid might actually produce a nerve gas–like substance to down-regulate its own nerve function, and then the nerve gas–detoxifying enzyme is part of that regulatory mechanism.

Just as the squid axon turned out to be extremely useful in neurological research, so also might this enzyme be helpful to mankind. "In a more immediately practical context," wrote Hoskin in 1990 (before sarin was actually used in Tokyo), "it seems reassuring that environmentally hazardous compounds such as nerve gases and insecticides may be detoxified by an enzyme found in the squid. Although such a relatively limited source is probably not practical for large-scale detoxifications, the potential exists for accomplishing this by a combination of genetic manipulation and biochemical engineering."

By What Name Shall We Call the Giant Squid?

T he mythological kraken has never appeared, but somewhat reduced in size, it has assumed a biological reality as *Architeuthis*. The earliest record seems to be of a carcass that was found in the Øresund between Malmö and Copenhagen in 1545. The contemporaneous Danish historian A. Sørensen Vedel (1542–1616), who was only three years old when the beast was found, later said that "a curious fish in monk-like shape caught in Øresund, was 4 ells long." (An ell was an old Danish measure that equaled 628 mm, so the creature was 8.16 feet long.) In his discussion of this event, Professor Steenstrup wrote, "Vedel could have had very good information about this event since it took place scarcely 40 years earlier; had it happened during his childhood he would have been able to obtain information from many people, then still alive, who had witnessed the event." In the "Annals of the Realm of Denmark," published in 1595, Arild Hvitfeld wrote, "In the year 1550 a curious fish was caught in the Øresund and taken to Copenhagen to the King; it had a head like a man and a tonsure on the head; it had a dress of scarlet like a monk's cloak."

The above quotes (and those that follow) are taken from a lecture given by Professor Steenstrup at the Danish Natural History Society on November 26, 1854. (His remarks were translated into English by Martina Roeleveld and Jørgen Knudsen, and

A lithograph of Professor Steenstrup in 1855, as he was beginning to publish his observations on Architeuthis.

published—fittingly—in *Steenstrupia,* the journal of the Zoological Museum of the University of Copenhagen.) Johannes Japetus Smith Steenstrup (1813–1897) was born in the province of Thy in northwestern Denmark. He studied medicine at the University of Copenhagen, but later switched to natural history. Described by Roeleveld and Knudsen as "extroverted, impulsive, choleric, willful, and restless," he became a lecturer in botany, mineralogy, and zoology and was eventually named a professor. Shortly after this appointment, Steenstrup became involved in a plan to combine the Royal Natural History Museum with the University Museum, and in 1867, the two museums were united as the Universitets Zoologiske Museum, and Steenstrup remained affiliated with it until his death. He published some 239 papers on subjects that ranged from peat bogs to the great auk, but he is probably best known for his work with cephalopods.

The name of Japetus Steenstrup is permanently connected with the name of *Architeuthis.* In the Linnaean system, the person who first bestows a name on a particular animal (or plant) has his name, and the date of the original publication in which the description first appeared, attached to the name. Thus, the official name of the genus to which all giant squid belong is *Architeuthis* Steenstrup 1856. There is no greater accolade in taxonomy than to have your name attached to that of an animal; it demonstrates that you were the first to differentiate that particular species from all the others. That Steenstrup's name is so closely affiliated with that of *Architeuthis* is surely the most persuasive testimony to his pioneering, innovative studies of the giant squid.

In his 1854 presentation, Steenstrup referred to an illustration of "a fish in the form of a monk," which appeared in Rondelet's 1554 *Libri de piscibus marinis* (Book of Marine Fishes), and purported to show the Sea Monk from Øresund, which is described as

> the animal which was caught in our time in Norway in a rough sea and which was immediately named a monk by all those who saw it. It appeared to have human features, but with a coarse and rude outline; the head was shaved and smooth; the shoulders were covered by a cape (or a cape with a cowl). It had two long fins instead of arms. The lower part ended in a broad tail.

Steenstrup then introduced Pierre Belon, another author who illustrated the monk-fish (in his 1551 *L'histoire naturelle des étranges poissons marins*), and added that "it lived for three days and did not produce a sound except for some deep sighs, indicating distress and sorrow." In neither Rondelet nor Belon does the illustration bear the slightest resemblance to a squid of any sort, but more like something that might be properly described as a Sea Monk.

Continuing within a historical framework, Steenstrup next brought in the Swiss physician Conrad Gesner (1516–1565), who wrote several volumes about animals, including *De piscium et aquatilium animatum natura*. Of particular interest to the Danish professor was a specific reference to Rondelet in this volume and the following description:

> The aquatic creature was caught in the Baltic, close to the town of Malmøe, which is situated four miles from Copenhagen, the capital of the Danish Kingdom. Head, neck, shoulders and breast: the head was shaven like that of a monk; hanging down over the neck, the shoulders and the breast was [something] resembling a monk's cape, and this was speckled with red and black spots. The cape passes downward into a skirt or such encircling tails with which we

To show the similarities between Rondelet's Sea Monk (left), Belon's Sea Monk (right), and the squid caught in the Kattegat in 1853 (center) Japetus Steenstrup lined them up in an 1854 paper. In order to give the squid a semblance of human arms, he folded its long tentacles under the body and had them reappear in a rather unusual fashion.

usually surround the hips and upper parts of the legs. Instead of arms and like-wise instead of [the] feet fins were present. The tail resembled a fish tail. The length of the monster was four ells. It was brought to the King and was preserved in a dried state as a rarity and a wonder. It was caught in the year 1546 A.D.

"I have not the slightest doubt," said Steenstrup, "that all the reports to which I have referred . . . deal with one and the same Sea Monk and the same capture of it. The events reported agree too well and are too unusual for any doubt to exist." But what was it? Steenstrup asked, "Could we, given these bits of information of how the Monk was conceived at that time, come so near to it that we could recognize to which of nature's creatures it should most probably be assigned?" And he answered, most emphatically, "The Sea Monk is firstly a cephalopod."

To make this point, he included a drawing of a standardized squid, with the various parts labeled and their functions explained.* As can be seen in the figure on page 64, Steenstrup had to strain to get the Sea Monk to look like a squid, and in order to explain the "arms," he said that the squid was lying on its two tentacles, which just happened to be arranged under it in such a way as to resemble arms. As for the face, that is a little more difficult, and he suggested that perhaps the ink colored that portion black, or that the skin was abraded and resembled a cowl. The "sighs" mentioned by Belon, of course, are the efforts made by the squid to breathe while out of the water. But in the end, having failed to show how the Sea Monk could actually have been a squid, Steenstrup said (to an 1855 audience), "Squids on the whole make a grim impression on all those who are not accustomed to seeing them more frequently. . . . Those animals aroused still greater astonishment in earlier times."

---

*For someone who probably saw most of his specimens in jars or spread out on a table, Steenstrup did a pretty good job of explaining teuthid behavior based on morphology—until he got to the long tentacles (labeled "b" in the drawing on page 64.) These arms, he wrote, "catch the prey by means of numerous suckers, or in certain cases, by sharp hooks with which they are provided . . . however, these arms are not only implements for catching, they also serve for locomotion, since the animal rows itself by means of them. Had it, under special conditions, to give up the swimming locomotion, which is its main way of movement that it performs with extraordinary speed, and be forced to proceed on a solid substrate, then the crawling is also done by means of these arms, the animal alternately fixing itself by different sets of suckers, carrying the body over the arms (a manner of locomotion that is found particularly in the eight-legged cephalopods)."

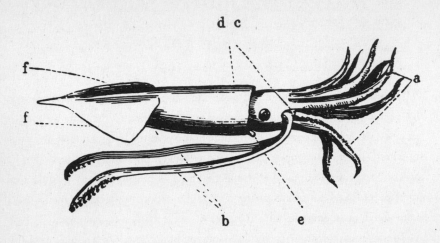

In his discussion of squid morphology, Steenstrup got most of the parts right, except for the long tentacles (labeled "b" in the drawing), which he correctly believed were used for catching prey, though he also opined that the animal used these appendages for rowing itself through the water.

"But is our Sea Monk a common squid?" asked Steenstrup. No, he answered. "Its amazing size . . . exceed[s] by several times that which naturalists are liable to assign to squids; indeed, they even consider almost all individual accounts with no less suspicion than the Sea Monk itself." By this time, Steenstrup had published his accounts of the "gigantic squids which were stranded on the coast of Iceland after storms in late summer 1639 and in December 1790," and at the lecture, he produced a jar that held the massive jaws of the giant squid that had been found at Aalbækstrand a year earlier—jaws the size of "a large clenched fist." This squid, he was told by wreck-master Kelder of Skagen, "was so large that according to fishermen it constituted a common cartload." (The Aalbækstrand specimen had washed ashore on the Øresund, but it was cut up for fish bait, and only the jaws, which measured 3 by 4 inches, were saved.) It was these jaws, even more than the early descriptions, that demonstrated to Steenstrup that there were monstrous squid in the sea:

> From all evidences the stranded animal must thus belong not only to the large, but to the really gigantic cephalopods, whose existence has on the whole been doubted. The few preserved remains, the horny jaws, are in the first place interesting because they are a tangible proof that huge cretures of this animal

class really inhabit the vast ocean. The supposition that only single individuals attain such unusual size seems unnatural and unfounded; it should rather be presumed that it is characteristic of the species that it attains such bodily development.

In the three hundred years that passed between the discovery of the Øresund Sea Monk and when Professor Steenstrup presented his ideas about its origins, many scholars had offered their opinions. But Messrs. Rondelet, Belon, and Gesner did not have the Aalbækstrand jaws in a jar, and did not know that there was any squid that could be 4 ells in length. Steenstrup concluded his lecture by showing the jar again, and said, "Had the horny beak been destroyed together with the remainder of the animal, the museum would presumably have been deprived of a very interesting specimen and our knowledge of the animals would have been a good deal less; but then neither would there have been an opportunity for all this talk about the Sea Monk." He could not make the case that his learned predecessors were actually describing a giant squid, but by 1854, he knew such a thing existed, and they never did.

He also had the benefit of numerous other reports of stranded creatures over the intervening centuries. After the Øresund Sea Monk, the next beaching occurred in 1639, at Thingøre Sand, in Iceland. In a paper that he read in 1849, Steenstrup quoted a 1639 description that appeared in *Annalar Björns å Skardsa* (in Danish), which was translated as follows:

> In the autumn on Thingøresand in Hunevandsyssel a peculiar creature or sea monster was stranded with length and thickness like those of a man; it had 7 tails and each of these measured approximately two ells. These tails were densely covered with a type of button, and the buttons looked as if there was an eye ball in each button, and round the eye ball an eyelid; these eyelids looked as if they were gilded. On this sea monster there was in addition *a single tail* which had grown out above those 7 tails; *it was extremely long, 4–5 fms* [7.5 to 9.4 m]; *no bone or cartilage were found in its body* but the whole to the sight and to the touch was like the soft body of the female lumpfish *(Cyclopterus lumpus)*. No trace was seen of the head, except the one aperture, or two, which were found behind the tails or at a short distance from them.

Even though the annalist managed to read the animal upside-down, confusing the head with the tail(s), it is quite obvious that the "monster" was a giant squid that had lost one of its arms and one of its tentacles.

Eggert Olafsen and Biarne Povelson were Danish travelers who journeyed through Iceland and wrote up their accounts in 1772.* They did not see a giant squid, but they did read the 1639 account of Björn of Skardsa in which the Thingøre Sand carcass was described. Olafsen wrote:

> It seemed to us immediately that the chronicler had made a mistake, confusing the forward and hind parts of the animal: in that case, the bud-studded appendices are not tails but arms or tentacles; and since only seven are mentioned, it would seem that the eighth had been torn off.
>
> Who cannot tell now that this animal was not just a very large squid—but of which species? That we cannot determine, since the structure of the stomach is not described, nor is that of the mouth, although it would appear that it had been damaged and distorted. The description of the buds and the cups is rather curious, but nevertheless carries an air of authenticity because of the precision with which some details, such as the colors, are described.

In 1673, a 19-footer was found on the beach of Dingle Bay, County Kerry, Ireland. It was displayed for all to see, and a broadsheet was printed up in Dublin, which announced the exhibition of:

> A Wonderful Fish or Beast that was lately killed, by James Steward, as it came of its own accord to Him out of the sea to the Shore, where he was alone on Hors-back at the Harbour's Mouth of Dingle-Icoush, which had two heads and Ten horns, and upon Eight of the said Horns about 800 Buttons or the resemblance of little Coronets; and in each of them a set of Teeth, the said Body was bigger than a Horse and was 19 Foot Long Horns and all, the great Head thereof Carried only the ten Horns and two very large Eyes, And the little head thereof carried a wonderful strange mouth and two Tongues in it, which had the

---

*The actual title of the book is *Vice-lovmand Eggert Olafsens og Lang-physici Biarne Povelson Reisen igiennen Island, foranstalter of Videnskabernes Saelskab i Kiobenhavn og beskreven of forbemeldte Eggert Olafsen,* or "Vice-lawman Eggert Olafsen's and physician Biarne Povelson's journey through Iceland, arranged by the Society of Sciences in Copenhagen and described by the aforesaid Eggert Olafsen."

natural power to draw itself out or into the Body as its own necessity required, there is several other very remarkable things to be observed in the said Monster, and in particular it had a Redish Coloured wrapper or Mantle growing and Sticking fast to the back thereof, and the Laps on both sides were loose, which was white within and Red without. Therefore all persons who desires to be further satisfied in the truth hereof, may see the said little head and two of the said Horn with the Coronets thereon, and a draft of the whole as it appeared altogether alive, with a certificate from responsible hands, and a real Relation of all the passages, witnessing the truth thereof, to their further admiration, at the Three Castles on the Lower end of Cork Hill.

The broadside is one of a number of documents that are included in an 1875 study by A. G. More, who takes the description and the accompanying drawing almost literally, and says: "I do not see why the extensible proboscis should not be accepted as correct, though the little eyes may have been added as ornaments by an enterprising showman." He then proposes to name the Kerry "monster" *Dinoteuthis proboscideus,* which can be roughly translated as "terrible squid with a big nose."* W. J. Rees, a biologist with the British Natural History Museum (and one of the editors of the Steenstrup papers), quoted this broadsheet in an article about giant squid that appeared in the *Illustrated London News* in 1949, and explained, "The 'little head,' of course, refers to the siphon through which water is pumped out to propel the squid through the water."

At irregular intervals, squid carcasses in various states of decomposition appeared on beaches around the world, and by 1735, Linnaeus included the creature he called *Sepia microcosmos* in the first edition of his *Systema naturae.* (Perhaps because he doubted its existence, Linnaeus dropped *S. microcosmos* from subsequent editions, and it does not appear at all in the definitive tenth edition, published in 1758.) Denys de Montfort then enters the story with his six-volume *Histoire naturelle générale et particulière des mollusques,* and although his *poulpe colossal* was a wild mixture of fact and fantasy, it encouraged other people to inves-

---

*More (an assistant naturalist in the Museum of the Royal Dublin Society) was acquainted with *Architeuthis,* as evidenced by his subsequent paper, "Some Account of the Giant Squid *(Architeuthis dux)* Lately Captured off Boffin Island, Connemara," which appeared in the same journal (the *Zoologist*) a month later.

tigate cephalopod stories. For example, Steenstrup found another early record of a giant squid in Iceland; in the winter of 1790, a creature that the people called *kolkrabbe* drifted ashore at Arnarnesvik.* This one seems to have been considerably larger than its predecessor, with a total length of 39 feet.

In Steenstrup's 1849 discussion, he included the following account of the animal's discovery:

> In November or December of last winter [1790] a creature drifted ashore in Arnarnæsvik here in the parish which people called *Kolkrabbe,* as according to them it completely resembled the animal called by that name in all features except the *unusual size, since the longest tentacula were more than 3 fathoms [ca. 5.60 m] long; but the body right from the head 3½ fathoms [about 6.60 m] long and so thick that a fullgrown man could hardly embrace it with his arms.* [The italics and bracketed equivalents are in the original.] It was intact when it was found, but I heard nothing about it until the whole animal was spoiled, cut to pieces, and as usual cured as bait for the cod, for which purpose the fishermen consider it excellent. The man who had had most to do with this creature only remembered that the long tentacula were four in number and all of them ten, just the number ascribed to that species of Sepia. According to Olafsen and Povelsen I think that there should only be two long tentacula. This animal presumably was the same as that peculiar animal at Thingøresanden in 1639, which the said travellers spoke about during their journey . . . and thus proves that it occurs in other places than in the Mediterranean, unless this is another and much larger species. *This animal had no bones except the well known one in the back.* Would not the Kolkrabbe deserve a closer examination than has hitherto been made?

Earlier authors, recognizing the necessity of naming those specimens of giant squid that occasionally washed up on shore, called them *Dinoteuthis proboscideus,* or *Sepia microcosmos,* or any other appellation that seemed descriptive and appro-

---

*According to a personal communication from Emil Ólafsson of Stockholm University, the Icelandic word *kolkrabbi* "is very much in use in Iceland and literally means 'coal-crab,' referring to the black defensive liquid of the beast and its [many-legged] resemblance to the crab." *Kolkrabbe* (or *kolkrabbi*) is sometimes used to describe an octopus, and *smokkur* or *smokkfiskur* is also used for squids. In *Zoology of Iceland,* Anton Bruun wrote that "*Todarodes sagittatus* [a common squid] is the *Kolkrabbi, Smokkur,* or *Smokkfiskur* of the Icelanders."

priate. In 1854, when he first described the "jaws of a giant cuttle-fish" that had washed ashore at Aalbækstrand in 1853, Steenstrup opined that they "must be supposed to approach the genus *Ommatostrophes* in form," a generic name that has not survived although it is now recognized as an earlier synonym of *Ommastrephes*. Steenstrup grouped this creature with the "huge cephalopods which were washed ashore in the years 1639 and 1791 on the north coast of Iceland," but he had no name for these animals (he also included the Øresund Sea Monk). It was three more years before he named the species *Architeuthis dux*.

Steenstrup described the beak of the Raabjerg specimen in 1855*, but he called it *Ommatostrophes pteropus*. In 1857, he renamed it *Architeuthis,* marking the official passage of the giant squid from the realm of fable into the scientific literature. This was the first appearance of the name *Architeuthis,* but because Steenstrup's findings were not submitted as printed papers but, rather, were recorded as he spoke at a meeting of Scandinavian naturalists, the rigorous requirements for the naming of a new species were not followed. *Architeuthis* made its debut thus: "On the supposition that the cuttle-fish washed ashore in 1853 was identical with that caught in 1546 or 1550 he had provisionally referred to its genus *Architeuthis* as *A? monachus* Stp." Also at that time, Steenstrup described another new species of giant squid, from the remains of another carcass that Captain Vilh. Hygom had brought from the Bahamas to Denmark. This animal "measuring six ells [377 cm =12.25 feet] overall with arms almost three ells long, [is] called *Architeuthis dux*."

Steenstrup described Captain Hygom's specimen as *A. dux,* but when A. E. Verrill saw the drawing of the lower jaw, he decided it was different enough from that of *A. dux* or *A. monachus* ("the beak is more rounded dorsally, less acute, and scarcely incurved, the notch is narrower, and the alar tooth is not prominent") that he would consider it a distinct species, which he designated (in an 1875 paper) *Architeuthis hartingii*.

---

*In his 1855 presentation, Steenstrup said "Raabjerg" (actually *Råbjerg* in Danish), but on the plate in which the jaws appear, he identifies the locality where the jaws washed ashore as "Aalbæk beach in the Kattegat." Raabjerg is a migrating sand dune, said to be the largest in Europe, on the Skagerrak (western) shore of the northern tip of the Jutland Peninsula, while Aalbækstrand is on the Kattegat (eastern) side. Given the confusion, we will probably never know where the type specimen of *Architeuthis* was found.

In his 1912 "Historical Review of the Genus Architeuthis" (in *The Cephalo-poda of the Plankton Expedition*), Georg Pfeffer (1854–1932) tried to clarify things when he wrote,

> In the year 1856, Steenstrup reported, at the meeting of the Scandinavian Naturalists at Christiana, among other subjects, on a large cephalopod stranded on the Danish coast in 1853, on the jaws of which he based the species *Architeuthis monachus;* and also on a specimen caught by Captain Hygom in the West Atlantic Ocean (north of the Bahamas) which he named *Architeuthis dux*. Steenstrup gives no diagnosis for the genus or the species. In 1858, the journal *Die Natur* published an article "On the Capture of a Merman in Oeresund in the Time of Christian III. by J. S. Steenstrup, after the lecture of the author, translated from Danish by H. Zeise." The name *Architeuthus* or *Architeuthis*[*] is not mentioned in the arti-cle, which is apparently a translation of Steenstrup's above-mentioned work of 1854. It therefore seems that the name *"Architeuthis monachus"* appeared only in the title of the original article. Consequently, *"Architeuthis monacus* Steenstrup 1854" was a nomen nudem; moreover, it was not based on the scientific exami-nation of the animal, but on the interpretation of traditional narratives and illus-trations of the 16th century. The name *Architeuthus monachus* therefore has no right to exist in the scientific nomenclature.

Although the genus is permanently associated with his name and he was the first scientist to describe a giant squid, because his presentations were often spoken and not transcribed, Japetus Steenstrup probably added as much confusion to the study of *Architeuthis* (or *Architeuthus*) as he did to explication. In summing up, Pfeffer wrote,

> At the meeting of the Scandinavian Naturalists in 1856, Steenstrup had already presented a plate depicting the jaws of *A. monachus*. Shortly there-

---

*Steenstrup also had a problem with the spelling of the *generic* name. In a footnote to his 1881 discussion of Verrill's paper on the cephalopods of the northeastern coasts of America, he wrote, "Verril [sic] and Tryon and other authors as well, persist in writing, and it seems delib-erately, Architeuthis instead of Architeuthus, which is twice wrong; in the first place because I did not call the genus by the first name but by the last one, which provided that it is not at variance with the fixed rules, should be respected; secondly, because the Greek, and Aristotles in particular, understood by '*Teuthis*' a smaller and weaker cephalopod, by '*Teuthos,*' however, a stronger and bigger one, and the composition *Architeuthis* thus becomes meaningless as if we in Danish would say giant-dwarf cephalopod or the like."

after printing of the text with plates apparently started for the production of an article that treated in greater detail the preliminary short discussion at the meeting. This work was not published in Steenstrup's lifetime; it first appeared in 1898; but since 1860 a portion of the text with plates was in the hands of teuthologists. Harting, Gervais and Verrill have mentioned this. Some (Harting, Verrill) corresponded with Steenstrup, Packard personally met him. As a result, Steenstrup's unpublished comments found a permanent place in literature with the same significance as if they had actually been published by himself. If only Steenstrup himself had been in possession of a diagnosis, be it only handwritten or part of his scientific conviction, for the genus *Architeuthis* or the two species *A. monachus* and *A. dux*, and if he had shared this with his correspondents, all statements in literature based on his private correspondence could be used as if they had been actually published. By this means, an arrangement could have been made according to the valid rules of priority. However, this was not the case, and so, by referring all findings on *Architeuthis* to species *A. monachus* and *A. dux*, which never existed, resulted the disorder in the literature, which together with the misinterpretations and errors of individual authors, can now be disentangled. But it seems best to quiet the matter in pointing out that any account of *A. monachus* and *A. dux* is pointless since these species do not exist, even after Steenstrup's work was posthumously published. We máy assume that they will not even exist when the excellent material in the Copenhagen Museum undergoes scientific revision; it can indeed be expected that the North Atlantic forms will be integrated into species diagnoses created by Verrill.[*]

Because of the early confusion, and also because there have been so few specimens available for examination, the separation of the species within the genus *Architeuthis* is still unsettled. In the 1984 *FAO Species Catalogue of the Cephalopods of the World*, Roper, Sweeney, and Nauen wrote, "Many species have been named

---

*Evidently, Steenstrup's curatorship was also not of the highest quality. In a 1983 discussion of the cephalopod collection in the Copenhagen museum, Kristensen and Knudsen wrote, "Steenstrup was restless, at times superficial, and not very interested in curatorial work. . . . A large part [of the collection] had to be discarded because the specimens were in very poor condition, owing to evaporation of the alcohol."

in the sole genus of the family *Architeuthis*, but they are so inadequately described and poorly understood that the systematics of the group is thoroughly confused." In other words, there might be a single species or there might be many. Because almost every known specimen of *Architeuthis* is a little different from every other—and many of them were described from only a beak or a piece of tentacle—almost every early description resulted in a new species.

The gladius (pen), mouthparts, penis, part of an arm, and several suckers of the specimen brought back from the Bahamas by Captain Hygom constitute the *type specimen* (the individual from which a new species was described) for the species named *Architeuthis dux* by Steenstrup in 1857. For the species he called *Architeuthis monachus*, all that remains of the specimen found at Aalbækstrand is the upper beak. The other seventeen (provisionally) recognized species are represented in various museum collections. (The place where the specimen was found and the date of the discovery are included):

A. *hartingii* (origin unknown, 1860)

A. *bouyeri* (Tenerife, 1862)

A. *harveyi* (Logy Bay, Newfoundland, 1873)

A. *princeps* (North Atlantic, 1875)

A. *proboscideus* (Dingle, 1875)

A. *sanctipauli* (St. Paul Island, Indian Ocean, 1875)

A. *mouchezi* (St. Paul Island, 1875)

A. *stocki* (New Zealand, 1880)

A. *verrilli* (New Zealand, 1880)

A. *martensii* (Japan, 1880)

A. *grandis* (Owen, 1881)*

A. *longimanus* (New Zealand, 1887)

A. *kirki* (New Zealand, 1887)

A. *physeteris* (vomited by sperm whale, 1900)

A. *japonica* (Japan, 1912)

---

*In his 1880 discussion of "some new and rare Cephalopoda," Owen examined a 9-foot-long arm in the collection of the British Museum (Natural History), which he decided belonged to a completely new species, and which he named *Plectoteuthis grandis,* from the Greek *plectos,* referring to a "well-defined fold between the rows of suckers." *(Plectoteuthis* is now recognized as a synonym of *Architeuthis.)* Either the location from which the arm had come was not recorded, or Owen thought it was unnecessary to include it.

The calmar gigantesque (Architeuthis sanctipauli) *that washed ashore on St. Paul Island in the southern Indian Ocean in 1875.*

A. clarkei (England, 1933)

A. nawaji (Bay of Biscay, 1935)

In 1984, before he co-authored the above disclaimer, Clyde Roper suggested that "the 19 nominal species can in fact be encompassed by only three: *Architeuthis sanctipauli* in the Southern Hemisphere, *A. japonica* in the northern Pacific, and *A. dux* in the northern Atlantic." And in 1991, refining it even further, Frederick Aldrich wrote, "I reject the concept of 20 separate species, and until that issue is resolved, I choose to place them all in synonymy with *Architeuthis dux* Steenstrup."

When Pfeffer attempted to summarize the genus *Architeuthis* in 1912, he wrote four pages on the various uses of *A. monachus, A. dux,* and *A. harveyi,* as well as *Megaloteuthis, Dinoteuthis, Plectoteuthis,* and even *Steenstrupia,* a genus that had been erected by T. W. Kirk in 1881. But after a thorough taxonomic analysis, he realized that the situation was still thoroughly muddled, and he wrote, "The question of whether several genera of *Architeuthis* exist cannot be determined either in the positive or the negative sense. . . . We have some very careful individual descriptions, but we scarcely know whether the characters given in the description reflect species-specific or merely individual features." The situation had not improved very much by 1982, when the Russian teuthologist Kir N. Nesis wrote (in his definitive *Cephalopods of the World*) that "15 nominal species have been described, but their differences have not been established, and the number of true species is unknown." And in the most recent analysis, Roper, Sweeney, and Nauen's 1984 FAO *Cephalopods of the World,* the authors wrote, "Many species have been named in the sole genus of the family *Architeuthis,* but they are so inadequately described and poorly understood that the systematics of the group is hopelessly confused."*

In a letter to me dated June 18, 1996, Martina Roeleveld of the South African Museum said, "So far, I have seen nothing to suggest that there might be more than one species of *Architeuthis.*" I believe that the taxonomy of *Architeuthis* is indeed "hopelessly confused," and it will not be resolved within these pages. (When quoting a primary source I have used the name as it appears in the original description.) No one truly knows how many species there are or what to call them. Until such time as a proper analysis can be done—something that would require far more specimens than are currently available—it is pointless to try to split the genus into so many species, and, at least for the moment, it makes sense to consider all specimens as *Architeuthis* sp.

Regardless of how many species there are, at least we know that we are discussing a very large, very poorly known squid when we speak of *Architeuthis.* It is the scientific name, after all, that is supposed to eliminate linguistic confusion; the giant squid is always *Architeuthis* in the scientific literature. In the vernacular, however, the names are different. The Japanese call this creature *dai-oo-ika,* which

---

*A good example of a "splitter"—one who separates the species on the basis of minor anatomical differences—is A. C. Stephen, who in 1962 wrote a paper for the *Proceedings of the Royal Society of Edinburgh* in which he meticulously identified the "North Atlantic" species (*A. harveyi, A. princeps, A. bouyeri, A. monachus, A. dux,* and *A. clarkei),* and explained how one could differentiate them. He was even able to identify the Scottish specimens on the basis of the shape of their tail fins.

can be translated as "great king of squid." (*Dai-oo* is "great king," and *ika* is the generic word for squid.) The Russian version is *gigantskiy kalmar;* the French, *céphalopode géant* or *encornet monstre.* The Spaniards would refer to it as *calamar gigante* or *megaluria;* the Portuguese say *lula gigante,* and the Italians, *calamaro gigante.* The Norwegians call the giant squid *kjempeblekksprut,* which can be literally translated as "giant" *(kjempe)* "ink" *(blaek)* "sprayer" *(sprutter)*—"giant ink sprayer."* (*Blekksprut* is the generic Norwegian word for octopus and squid.) The Danes use a similar name—*kaempeblaeksprutte*—but the Swedes call it *kämpe-bläkfisk,* or "giant ink fish." In Germany it is also "giant ink fish," but there the translation is *Riesentintenfisch,* or *Riesenkalmar.*

---

*In my 1995 book *Monsters of the Sea,* I managed to misspell the Norwegian name consistently. Somehow, I managed to insert an extra syllable, which turned *kjempeblekksprut* into *Kjempebleblekkspruten.* The addition of "*ble*" turned "giant ink sprayer" into "giant *diaper* ink sprayer."

Architeuthis
Appears

New England sperm whalers often noticed that the whales vomited up some sort of "arms" while in their death throes, and while their biology was often less than reliable, there was no question that they were indeed seeing pieces of the kraken. Here is whaleman Charles Nordhoff (the grandfather of the Charles Nordhoff who cowrote *Mutiny on the Bounty*) describing the animal in his *Whaling and Fishing*, published in 1856:

> Sperm whales feed on an animal known among whalemen as "squid," but which is, I believe, a monster species of cuttle fish. These, like their smaller congeners, cling to the rocks, the larger species, of course, having their haunts at the bottom of the sea, while the smaller frequent only the shores of bays.
>
> Very few men have ever seen an entire squid or a sperm whale cuttle-fish[*], and I incline to the belief that most of the few instances on record of their appearance at the surface, are apocryphal. Whalemen believe them to be much larger than the largest whale, even exceeding in size the hull of a large vessel; and those who pretend to have been favored with a sight of the body, describe it as a huge, shapeless, jelly-like mass, of a dirty yellow, and having on all sides of it long arms, or feelers, precisely like the common rock-squid.

In the North Atlantic, *Architeuthis* was about to make its appearance. On November 30, 1861, off the island of Tenerife in the Canaries, the lookout of the French steam warship *Alecton* (Lieutenant Bouyer commanding) spotted a monstrous animal floating on the surface, with a brick-red body that was 16 to 18 feet long (not counting the tentacles) and glimmering green eyes, which made the crew uneasy. As the warship approached, the creature tried to move out of the way, but it did not dive below the surface. Lieutenant Bouyer determined to secure the animal, but he was reluctant to lower the boats for fear of the harm

---

*Although it looks like something is missing from the name, this is the way it appears in the original. I assume that Nordhoff's "sperm whale cuttle-fish" is that species of cuttlefish that he believes is eaten by the whale.

The French warship Alecton approaches an enormous poulpe on the surface off Tenerife.

that might come to his men. When they began shooting at it, the animal dived, but it always reappeared. One shot hit a vital organ, for the animal vomited blood and froth, and at this time, it emitted a strong odor of musk. When they threw a loop around it to haul it aboard, the rope cut through the body, and the head and tentacles fell into the sea and sank. They brought the remaining tail section to Tenerife.

There Lieutenant Bouyer filed a report, which was forwarded to Sabine Berthelot, the French consul, who presented the paper at the December meeting of the French Academy of Sciences.* From the similarity of the details, it appears obvious that Jules Verne saw the Bouyer report and adapted it for *Twenty Thousand Leagues Under the Sea*, but made the *poulpe* far more aggressive than it actually was. (In one instance, Bouyer wrote, "*Sa bouche, ou bec de perroquet, pouvait offrir près d'un demi-mètre*"—"its mouth, like the beak of a parakeet, could open nearly a half meter," and Verne's version, employing the same device, reads,

---

*The French malacologists Henri Crosse and Paul Fischer analyzed the reports of the *Alecton*, and in 1862 they published "Nouveaux documents sur les céphalopodes gigantesques," in which they suggested that the gigantic squid was only an overgrown *Loligo*, which, unlike the higher vertebrates, could continue to grow throughout its life.

"*un bec de corne fait comme le bec d'un perroquet*"—"a horny beak like the beak of a parakeet.")

The first giant squid ever taken in American waters was collected in October 1871, by fisherman in the Gloucester schooner *B. D. Haskins* on the Grand Banks. To get an idea of how the fishermen must have felt upon seeing a giant squid floating in the water, here is the text of a letter sent by Mr. James G. Tarr of Gloucester to A. S. Packard, the founder and editor in chief of the *American Naturalist*:

> The weather being fine and pleasant [the captain] ordered the boat lowered, and sent two men to learn what it might be; they returned reporting the object to be a mass of floating jelly, or something unknown to them. The Captain then with hooks and gaffs and more men went to investigate; he found it quite dead, each end hanging under water, only the centre on the surface. After towing it

Illustrator N. C. Wyeth painted Adventure of the Giant Squid of Chain Tickle *in 1940 to illustrate Norman Duncan's short story of the same name.*

alongside the schooner, he took his purchase or halyards to hoist the monster out of the water, and on seeing its head, declared it to be that of a squid, saying he had heard of squid that size, but never saw the like before. After it was got on board, the second hand or mate informs me he measured the body with a rule and found it fifteen feet long, four feet eight inches round. The long arms were badly eaten; judged they might be nine or ten feet long; two were shorter than the former, perhaps two or three feet; did not measure the arms but judged them twenty-two inches round, also judged the weight to be two thousand pounds, and would fill eight or ten barrels.

Packard (who reproduced the letter in an article entitled "Colossal Cuttle-fishes") sent a photograph of the squid to Professor Steenstrup in Copenhagen, who declared that it was indeed *Architeuthis monachus*.

In the fall of the following year, an even larger specimen washed ashore at Coomb's Cove, Fortune Bay, Newfoundland, where it was secured by local fishermen. One tentacle was about the diameter of a man's wrist, and was 42 feet long. The body was 10 feet long (and "nearly as large round as a hogshead"), so the total length was 52 feet.

Addison Emery Verrill (1839–1926) of Yale University, who described most of the specimens of Architeuthis *that washed ashore in Newfoundland between 1871 and 1881. With J. H. Emerton, he also designed the first of the life-sized giant squid models.*

In late October 1873, at Portugal Cove, in Conception Bay, New-foundland, a giant squid attacked a small boat. Two of the arms were chopped off and brought ashore, and eventually brought to the attention of the Reverend Moses Harvey of St. John's. (Portugal Cove is due west of St. John's, about ten miles across the peninsula.) Depending upon whose account you read, there was either one man in the boat, two men, or two men and a boy (see pages 82–84).

This event was fictionalized in a 1940 children's story by Norman Duncan entitled "The Adventure of the Giant Squid of Chain Tickle." In Newfoundland waters, "Billy Topsail" and "Bobby Lot" battle a giant squid that they eventually subdue. They present the carcass to "Dr. Marvey" (Moses Harvey) and it is described in a monograph by "Professor John Adams Wright" (Verrill). (The story appears in the *Anthology of Children's Literature,* where the N. C. Wyeth illustration on page 80 was originally published.) In 1995, Don Reed wrote *The Kraken,* a "young adult" book about this episode (discussed on page 187), in which young Tom Piccot's heroic actions are also emphasized. Although the story of the brave lad has become a permanent part of Newfoundland and *Architeuthis* folklore, there is a strong possibility that the youngster never existed, or if he did, that he had nothing to do with the giant squid.

At Harvard College, Addison Emery Verrill (1839–1926) studied under Louis Agassiz, and upon graduation in 1862, he was appointed professor of zoology at Yale, a post he held until he retired in 1907. From 1872 to 1880, he made a special study of the giant cephalopods from the North Atlantic, especially Newfoundland, and managed to secure several specimens for the collection at New Haven. (With his draftsman, J. H. Emerton, he was also responsible for the first life-sized models of the giant squid, as discussed in the chapter "The Models of *Architeuthis.*") In 1873, when the Reverend Moses Harvey obtained the tentacle of the Portugal Cove specimen, he decided that the interests of science would best be served by sending it to Verrill, who incorporated it into his voluminous studies of the giant squid of eastern North America. Obviously intrigued by these "gigantic cephalopods," Verrill examined, dissected, and described them with unmitigated delectation. From 1874 to 1882, he wrote no fewer than twenty-nine scientific papers in which *Architeuthis* was discussed.

In his 1875 discussion of the "colossal cephalopods of the Western Atlantic," he acknowledges "Mr. Alexander Murray, Provincial Geologist, who cooperated with Mr. Harvey in the examination and preservation of these specimens, and who has also written some of the accounts of them that have been published." Murray published two accounts: the first in 1874 in the *Proceedings of the Boston Society of Natural History,* and the second in the same year in *American Naturalist.* Both accounts are transcripts of a letter dated November 10, 1873, that Murray sent from St. John's, Newfoundland, to Professor Louis Agassiz in Cambridge, Massachusetts. "On or about the 25th of October last," wrote Murray, "while a man by the name of Theophilus Piccot was engaged at his usual occupation of fishing . . . his attention was attracted to an object floating on the surface of the water, which at a distance he supposed to be a sail, or the *débris* of some wreck, but which proved on nearer inspection to be endowed with life." It was a giant squid, and after he struck it with a boat hook, it attacked his boat, and "immediately afterward, threw its monstrous tentacles over the boat, *had not Piccot with great presence of mind severed one (or more) of the tentacles with his axe.*" (My italics.) There is not a word about another man in Murray's account, let alone a twelve-year-old boy.

Moses Harvey (1820–1901) was the rector of St. Andrews Free Presbyterian Church in St. John's, and a prolific lecturer and writer in the community. He wrote some nine hundred articles for the *Montreal Gazette* alone, which he signed "Delta." His work also appeared in *American Naturalist* and the *Annals and Magazine of Natural History,* and he was the author of *The Textbook of Newfoundland History* (1890) and *Newfoundland, the Oldest British Colony* (1893). His writing skills may not have been limited to historical exposition.

Harvey sent a letter on November 12, 1873, three weeks after the event, which was published early the following year in the British journal *Annals and Magazine of Natural History.* He described the battle with the giant cuttlefish of Conception Bay: "Two fishermen were out in a small punt. . . . Observing some object floating in the water, they rowed toward it, supposing it to be a large sail or the *débris* of a wreck. On reaching it, one of the men struck it with his 'gaff,' when immediately it showed signs of life, reared a parrot-like beak which they declare was 'as big as a six-gallon keg,' with which it struck the boat violently. . . . One of the men seized a small axe and severed both of the arms as they lay over the gunwale of the boat, whereupon the fish moved off and ejected an immense amount of inky fluid which darkened the water for two or three hundred yards."

Twenty-five years later, when the Reverend Harvey told the story again ("How I Discovered the Great Devil-Fish"), Daniel Squires and Theophilus Piccot have acquired an apprentice in the form of twelve-year-old Tom

Piccot, the son of Theophilus. In Harvey's retelling, it is young Tom who saves the day:

> The terrible monster then disappeared beneath the surface, dragging men and boat with it. The terror-stricken fishermen were completely paralyzed and thought their last hour had come. The water was pouring into the boat as it sank lower and lower, and in a few seconds all would have been over with the unfortunate men. Quick as lightning, however, the boy Piccot took in the situation, and seizing a small tomahawk, that fortunately lay in the bottom of the boat, the brave little fellow dashed forward, and with two or three quick blows cut off both arms as they lay over the edge of the boat.

It appears that young Tom Piccot, immortalized in subsequent renditions of the story, might only have been a literary invention of Moses Harvey's. (When Annie Proulx tells the story in her Pulitzer Prize–winning novel *Shipping News,* there are only two men.) In his 1899 article, Moses Harvey may have been defending his invention when he wrote, "I lost sight of Tom for twelve or fourteen years, and when I next saw him he had grown into a handsome, strapping young fisherman. I also learned that he was known for his daring, and not less for his kindly, generous disposition."*

Ultimately, it does not matter if the incident involved two men or three, for it was the tentacle, some 19 feet long, that proved to be the most scientifically exciting. (Professor Steenstrup's 1857 description was based on the jaws, gladius, part of an arm, and several suckers; and when the crew of the *Alecton* tried to haul in the carcass in 1861, they lost everything but the tail.) After the arms were cut off, the squid emitted a cloud of ink and returned to the depths. In Portugal Cove, somebody carried home the longer arm and another shorter piece, which was thrown to the dogs. The tentacle was brought to the Reverend Harvey, who placed it in the St. John's Museum and had it photographed. Despite his occasional novelistic pretensions, the Reverend Harvey will probably be remembered best for being the recipient of the arm of a giant squid in 1873. As he wrote (in his 1899 article), "I was now the possessor of one of the rarest curiosities in the

---

*The date of this event seems also to be a variable. Murray (1874) says it was "on or about the 25th"; Verrill (1879) says it was October 27; and the Reverend Harvey—in 1874 and 1899—says it was the 26th.

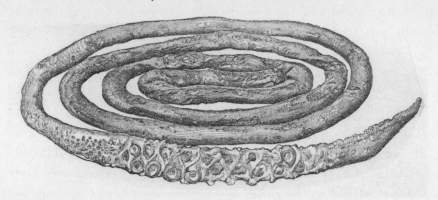

The 19-foot tentacle of the first Architeuthis *ever examined on land. It was hacked off a living animal in October 1873 by a young Newfoundland fisherman, who brought it to the attention of the Reverend Moses Harvey.*

whole animal kingdom—the veritable tentacle of the hitherto mythical devilfish, about whose existence naturalists had been disputing for centuries. I knew that I held in my hand the key of the great mystery, and that a new chapter would now be added to Natural History."

Also in Newfoundland, in December 1873—only a month after the incident at Portugal Cove—four fishermen were hauling in their herring net in Logy Bay when they realized that the heavy, writhing net contained more than herring. A giant squid was trapped in the net, and the fishermen managed to kill it with their knives. They too brought the carcass to the Reverend Harvey, who would later write, "I remember to this day how I stood on the shore of Logy Bay, gazing on the dead giant. . . . I resolved that only the interests of science should be considered. I speedily completed my bargain with the fishermen, whom I astonished by offering 10 dollars to deliver the beast carefully to my house." To enable the steady stream of curious visitors to see the carcass, Harvey draped the head and arms over his sponge bath. He then had it photographed by Mr. John Maunder, and wrote, "The photograph, like George Washington, cannot tell a lie. Had I published the story without the attesting photographs, I have little doubt that I should have been pronounced a Munchausen on a small scale, and my story would have been placed on the same level as those of the mythical sea-serpent." The photograph shows the arms and the beak of the squid (the body was lost when they cut off the head) draped over the sponge bath on a patterned carpet. It is not known what Mrs. Harvey thought about the cephalopod arrangement in her living room. The tentacles were 24 feet long and the entire animal was 32 feet long.

The Reverend Harvey sent the specimen to New Haven, where Verrill "made an exhaustive study of it, and described and figured it in numerous scientific periodicals; so that its fame spread over the entire world." As a tribute to his friend, correspondent, and supplier, Verrill named the species *Architeuthis harveyi,** but "another eminent English naturalist, Dr. W. Saville-Kent, F.L.S., F.Z.S., called it *Megaloteuthis Harveyi* in recognition of the great service to science rendered through Mr. Harvey's steps taken to preserve these valuable specimens." Harvey tells us that he "endeavoured to bear these honours meekly," but it must have been difficult. P. T. Barnum wrote to Harvey and ordered "two of the very largest devil-fish, and to spare no expense." "He probably thought they were as plentiful as cod-fish," wrote Harvey. A friend of his congratulated him on the "prospect of going down to posterity mounted on the back of a devil-fish!"

During the 1870s in Newfoundland—for reasons that probably had to do with some unexplained climatic or oceanographic changes—dozens of giant squid washed ashore or were seen floating at the surface. According to Verrill's analysis, some fifty or sixty were collected by fishing vessels on the Grand Banks and used as bait for cod or as dog food. Another twenty-three were the subjects of Dr. Verrill's meticulous examination.

Verrill, who had an iron constitution and an almost limitless capacity for work, began publishing papers on these specimens almost as fast as they came in. In an 1879 paper entitled "The Cephalopods of the North-Eastern Coast of America. Part I: The Gigantic Squids *(Architeuthis)* and Their Allies; with Observations on Similar Large Species from Foreign Localities," Verrill listed every specimen of *Architeuthis* known at the time of publication. That such a thing was possible is a direct function of the rarity of this animal; few other large creatures are known from such a small number of individuals. (Since Verrill's day, of course, the total has increased, but each giant squid that washes up or is taken from the stomach of a sperm whale is still an occasion for a teuthological cele-

---

*In zoological nomenclature, the generic name (*Architeuthis,* in this case) is supposed to be capitalized, while the specific name *(harveyi)* is not. In quoting from Harvey's article, I have retained his style.

In December 1873, while hauling in their herring nets, fishermen in Logy Bay, Newfoundland, found that they had trapped a giant squid. After killing it with their knives, they lost the body, but brought the head and tentacles to the Reverend Moses Harvey, who draped it over a sponge bath in his living room. At the top is the squid's beak.

bration.) Verrill celebrated what was one of the most remarkable and unexpected arrivals in the history of zoology, crypto- or otherwise. An animal that was believed by many to be mythological verified its corporeal existence by appearing all over the beaches and shallows of Newfoundland from 1871 to 1881.

In 1872, the Reverend Harvey wrote to Verrill that he had been apprised of a large specimen that was cast ashore at Bonavista Bay. Only the jaws and tentacular suckers were measured—the largest ones were 2.5 inches in diameter—but Harvey's informant, a fellow clergyman named Munn, remembered that the short arms, about 10 feet in length, were "thicker than a man's thigh." The Logy Bay animal filled the quota for 1873, and in 1874 another giant came ashore in Newfoundland, this one at Grand Bank, Fortune Bay. The magistrate of Grand Bank, a Mr. George Simms, had examined the carcass before it was cut up for dog meat, and had

A. E. Verrill's reconstruction of "Architeuthis Harveyi," the giant squid that washed ashore in Logy Bay, Newfoundland, in December 1873 and was brought (in pieces) to the Reverend Moses Harvey in St. John's.

recorded that the longest tentacles were 26 feet long and 16 inches in circumference. The body was 10 feet long, making the Fortune Bay specimen a 36-footer.

Another animal washed ashore at Harbour Grace during the winter of 1874–75, but, reported Verrill, "it was destroyed before its value was known, and no measurements were taken." In September 1877, however, a "nearly perfect specimen" came ashore at Catalina, Trinity Bay, during a severe gale. It was living when found, so it was exhibited for two or three days in St. John's, and then packed in brine and exhibited at the New York Aquarium. Based on an examination of the suckers (which were sent to him by the proprietors of the aquarium), Verrill identified it as *Architeuthis princeps*. As of that time, it was "the largest and best specimen ever preserved." The body was 9.5 feet long, and the longest tentacle was 30 feet long. The eyes were 8 inches in diameter.

Although neither Verrill nor his trusty informant saw it, Verrill quotes the report of a Dr. D. Honeyman of Halifax, Nova Scotia, who recorded the state-

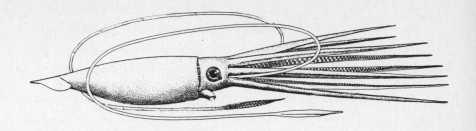

J. H. Emerton's drawing for A. E. Verrill's 1879 description of the giant squid that washed ashore at Catalina, Trinity Bay, Newfoundland, on September 24, 1877.

ment of a gentleman who was present at another capture, this one in the Strait of Belle Isle. (No date is given.) This giant squid, which was 52 feet long including the body and longest arms, was disturbed with an oar as it lay peacefully in the water, but Verrill writes that it was "probably disabled [because] animals of this sort probably never float or lie quietly at the surface when in good health." In 1876 a specimen was discovered at Hammer Cove, on the southwest arm of Green Bay, in Notre Dame Bay, Newfoundland. By the time it could be examined, it had been partially devoured by foxes and birds, so only a 5-foot hunk of the body remained, with 2-foot stumps attached to it. In 1877, Harvey wrote to Verrill describing a specimen that was stranded at Lance Cove, Trinity Bay, some twenty miles up the bay from the location of the Catalina stranding. This animal was alive and thrashing when it was found, and in its struggles to get back into the water, "it ploughed up a trench or furrow about thirty feet long and of considerable depth by the stream of water that it ejected with great force from its siphon. When the tide receded it died."

In a letter to the *Boston Traveller*, dated January 30, 1879, the Reverend Harvey described a huge squid that had been sighted off Thimble Tickle the previous November. Three fishermen spied

> some bulky object, and, supposing it might be part of a wreck, they rowed toward it, and, to their horror, found themselves close to a huge fish, having large glassy eyes, which was making desperate efforts to escape, and churning

the water into foam by the motion of its immense arms and tail. It was aground and the tide was ebbing. From the funnel at the back of its head it was ejecting large volumes of water, this being its method of moving backward, the force of the stream, by the reaction of the surrounding medium, driving it in the required direction. At times the water from the siphon was as black as ink.

The intrepid fishermen threw a grapnel at it, dragged it to shore, and tied the line to a tree. The squid died and was cut up for dog food, but the Reverend Harvey, the happy coordinator of all this information, estimated that the Thimble Tickle squid was 55 feet long.

Several more giants appeared on Newfoundland beaches before the *Architeuthis* Decade came to a close: in November 1878, "a fine and complete specimen" was captured at James's Cove, Bonavista Bay. The fishermen, "as usual, indulged immediately in their propensity to cut and destroy," so none of it was preserved. In a later letter from an observer, however, its capture was described: "One of the men struck at it with an oar, and it immediately struck for shore and went quite upon the beach." The body was 9 feet long and the tentacles were 29 feet each. On December 2, 1878, after a heavy gale, a specimen identified as *A. princeps* came ashore at Three Arms; its body was measured at 15 feet from the beak to the end of the tail.

The decade closed with no more strandings in Newfoundland, but in 1881, at Portugal Cove, one last specimen was found floating at the surface. It was packed in ice and shipped to New York by steamer, where the indefatigable Professor Verrill journeyed from New Haven and examined it at the museum of Mr. E. M. Worth, at 101 Bowery. It was too damaged for accurate measurements, but Verrill believed that the total length was about 20 feet. He wrote, "The color, which is partially preserved, especially on the arms and on the ventral surface of the body . . . consist[s] of small purplish-brown chromatophores more or less thickly scattered over the surface."

In April 1875, three fishermen were at sea in a curragh near Boffin Island, off the Connemara coast of Ireland. Upon noticing a large, shapeless mass floating at the surface they rowed over to investigate, and were more than a little surprised to find that the mass was a giant squid. They grabbed one of the tentacles and lopped it off, causing the creature to erupt in a violent flurry of foam and ink.

They amputated one arm at a time—always trying to keep out of range of the remaining flailing appendages—until they had a feebly thrashing, moribund, almost armless squid that they towed to shore. Describing the events attendant upon the securing of this beast, A. G. More wrote:

> The history lately given in a newspaper, the "Galway Express"—which has also been published in the *Zoologist* for June—is no myth, and the great size of the animal is sufficiently proved both by letters received from Sergeant O'Connor of the Royal Irish Constabulary stationed in Boffin Island, and by the portions of this great squid which he has sent up to Dublin. Though imperfect, both tentacles and arms are represented, and the huge beak, about five inches across, is now to be seen in the Museum of the Royal Dublin Society.
>
> The animal was killed on the 25th of April; and, as the men who attacked it were in a small boat, they could only bring ashore the head and some of the arms—viz., the tentacles and two of the short arms. The head and eyes were unfortunately destroyed, but Sergeant O'Connor managed to rescue, and has transmitted to us, the greater part of both tentacles, one short arm, and the beak. He measured the tentacles when fresh as reaching to the length of thirty feet, and the portions of them which we have received—shrunk and distorted as they now are—still measure fourteen and seventeen feet, when the pieces are put together.

Something must have been happening in the icy depths of the oceans during the years 1870 to 1890 that caused giant squid to appear at the surface and on certain beaches, but unless there is a repeat of this inexplicable phenomenon, we might never know what it was. A. E. Verrill was as perplexed as everyone else, and wrote (in 1881),

> The cause of so great a mortality among these great Cephalopods can only be conjectured. It may have been due to some disease epidemic among them, or to an unusual prevalence of deadly parasites or other enemies. It is worth while, however, to recall the fact that these were observed at the same time, in autumn, when most of the specimens have been found cast ashore at Newfoundland, in different years. The season may, perhaps be just subsequent to their season for

reproduction, when they would be so much weakened as to be more easily over-powered by parasites, disease, or other unfavorable conditions.

Eighty-seven years later, with more data and more specimens available, Frederick Aldrich was able to speculate that fluctuations in the Labrador Current were responsible for the appearance of giant squid off Newfoundland every ninety years or so. When the cold portion known as the Avalon Branch hits northeast-ern Newfoundland, the squid come close to shore, following the cold mass of water. He predicted that the data would show that the next period of *Architeuthis* strandings would occur around 1960, and he was proven correct when nine spec-imens stranded between 1961 and 1968. (That the strandings were not related to a reproductive cycle was demonstrated by the sexual immaturity of all nine spec-imens.) According to Aldrich's predictions, the next appearance of giant squid in Newfoundland ought to occur around the year 2050.

This might explain the Newfoundland strandings, but what are we to make of a similar, albeit smaller, invasion of *Architeuthis,* halfway around the world in New Zealand, during the same years? In May 1879, a specimen was stranded at Lyall Bay, Cook Strait, New Zealand, and described by T. W. Kirk to the Wellington Philosophical Society. It took three years before the specimen was examined by scientists, and by that time, all that remained was the beak, radula, and a few suckers. A year later, another giant squid stranded at Island Bay (also in Cook Strait), and again Kirk described it, naming it *A. verrilli.* In their 1982 article in *Scientific American,* Clyde Roper and Kenneth Boss called the 1880 Island Bay specimen, at 55 feet long, "the largest specimen recorded in the scientific literature." In 1886, a specimen was found on the beach at Cape Campbell, and the author of its record, a lighthouse keeper named C. W. Robson (not to be confused with G. C. Robson, a teuthologist at the British Museum), named it *A. kirki,* for T. W. Kirk. Then in 1887, another squid came ashore at Lyall Bay, and Kirk wrote, "And now we have another of these highly interesting, but very objectionable, visitors." Kirk proceeded to the beach, made a careful examination, took notes, measurements, and also "obtained a sketch, which, although the terribly heavy rain and driving southerly wind rendered it impossi-ble to do justice to the subject, will, I trust, convey to you some idea of the gen-eral outline of this most recently arrived Devil-fish." Kirk named this new species

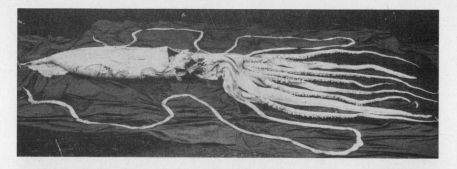

Measuring 32 feet from tentacle tip to tail, this specimen of Architeuthis was collected in 1896 at Hevne, outside Trondheimsfjord in Norway.

A. longimanus for the enormous length of its arms: a local fisherman named Smith had paced it off at 62 feet, but Kirk measured it at 55 feet, 2 inches, "or more than half as long again as the largest species yet recorded from these seas." Although the total length of the Lyall Bay *longimanus* was more than 55 feet, making it the longest of all measured giant squid, its body (mantle) was only 5 feet, 7 inches long.

Of these nineteenth-century New Zealand strandings, R. K. Dell wrote (in 1970), "Most specimens have been in poor condition or have not been examined by competent workers. Strandings of these squids on uninhabited stretches of our coasts and sightings of dead bodies at sea are probably much more common than the published records would indicate." When a giant squid washed ashore at Makara in 1956, it gave Dell the opportunity to examine a "perfectly fresh" specimen. The tentacular arms had been sheared off obliquely about a meter from the head (possibly by a sperm whale), but the body was in good condition. Where the color had not been abraded, it was described as being somewhere between madder brown and brick red. In recent years, with the advent of deep-water trawling, so many specimens of *Architeuthis* have been captured in New Zealand and adjacent waters that it is becoming clear that this may very well be the richest area in the world for giant squid. (See pages 102–104 for a discussion of the recent New Zealand occurrences.)

Starting in this century, the geographic reach of recorded beachings has grown more widespread. In March 1909, Massachusetts recorded its first giant squid. (The second would not appear for another seventy-one years.) Off the Cape Cod village of Truro, the fishing schooner *Annie Perry* found a giant squid floating on the surface. In an anonymous letter written to Henry Blake (who quoted it in an article in the malacological journal *Nautilus*),

> It was perfectly fresh, and the crew took some of it for bait and caught quite a number of fish. I saw one of the tentacles . . . and it was seven feet six inches long, and the suckers were as large as a silver quarter. A piece of the body was, I should think, four inches in thickness, and the tentacles must have been four inches in diameter at the larger end. . . . The captain of the vessel who took the squid says it was a very little larger than their dory, which is 16 or 17 feet long. . . . The whole body was about as large around as a fish barrel.

Norway, a country that is almost all west coast, has also been a fertile location for observing dead or dying giant squid. In 1916, August Brinkmann listed some of the strandings prior to the year of his publication, but he devoted most of his attention to an animal that was found swimming at the surface at Austrheim, just north of Bergen, in November 1915:

> Joakim Lerøen . . . discovered the animal swimming on the surface of the water, criss-crossing its way towards the neck of the bay, with its rear-end in front. He ran into the boat-house and grabbed a gaff with which he cut into the eye of the squid when it resurfaced at the breath of the bay. Up to this point, the animal appeared to be light-colored but when the gaff cut into its eye, it turned deep purple. It simultaneously released its ink bag, thus darkening the surrounding water, and threw one of its giant tentacles above the surface, attempting twice to clasp the man. Meanwhile, the fisherman's neighbor had come to

A giant squid that washed ashore at Ranheim, Norway, in October 1954, spread out on the ground to show its size and the arrangement of arms and tentacles.

help, and the two men managed to pull the animal up and fasten it in the shallow water.

The following year, a living squid was seen on the beach at Hellandsjø; like the Austrheim squid, it too was in the Hevnefjord. A lad approached the squid and put his foot on one of the arms. With its suckers, the squid fastened so firmly to his shoe that he had to pull his foot out to get away. Fishermen killed the squid and cut it up for bait, and Otto Nordgård, who wrote up the report in 1923, got "only some remnants," including the 20-foot tentacles.

In another Norwegian discussion of a stranded individual (this one a 30-footer, at Ranheim in 1954), Erling Sivertsen includes a map of Norway, showing the locations and dates of eighteen strandings that occurred from 1874 to 1954. One might envision a central location in the North Atlantic with lines radiating outward toward Newfoundland in the west and Norway in the east, whence the giant squid disperse, but such a concept would require far more information about the distribution and habits of *Architeuthis* than is actually available. (Although several of the Norwegian records occurred during the decade from 1870 to 1880, the greater proportion occurred afterward, showing what would appear to be an irregular chronology, with a couple of records during every decade from 1900 to 1950.)

Architeuthis continued to show up in Newfoundland, although now with much less frequency. In 1933 and 1935, two more appeared. The 1933 specimen washed ashore in December, and as might be imagined, this is not the best month in Newfoundland to examine objects on the beach, and it was not until February that it was examined. It was kept in cold storage, so that the internal organs were well preserved, but, as Nancy Frost wrote in her 1934 report, "the head and tail were detached from the body, and all the arms, with the exception of one almost intact tentacular club, . . . were hopelessly mutilated."

In a newspaper interview published in 1996, Reuben Reid, who was twelve years old when he and Richard Gosse found the carcass, said that he "saw an object floating in the water . . . as it drifted closer, I discovered it was a squid, a giant squid." It had been pushed close to shore by a storm, and when Reid tried to get a rope around it, the head was severed. They loaded the head and body on a sled and dragged it back to town, where it was weighed and measured. "We stuck the head back on," said Reid. "Its eyes were like small saucers, the body was about 10 feet long with the longest tentacle being 12 feet long, the tail two feet wide. It weighed 570 pounds."*

Across the Atlantic, in January 1933, W. J. Clarke "received a message that a large Shark had washed ashore in the South Bay at Scarborough [Yorkshire]. On the way down to view it, it grew into a Whale, but on arrival the mysterious creature was found to be an immense squid." Damaged by people who "danced on the body, scrubbing off the delicate outer skin until only the vestiges remained," the carcass was transported to the British Museum (Natural History), where it was described by that institution's renowned teuthologist, G. C. Robson. In his

---

*The interview, written by Lillian Simmons and published in the Newfoundland *Compass* for April 2, 1996, concerned Reid's discovery of the squid carcass in the ice, and also the creation of a full-sized model based on this specimen that was built in Dildo as a tourist attraction. This is one of the models discussed in more detail later in "The Models of *Architeuthis*."

1933 description, Robson named it *Architeuthis clarkei* "in honour of Mr. W. J. Clarke of Scarborough, who has for many years been a zealous observer of Cephalopods in the North Sea, and to whose activity I am indebted."

Robson's writings on the Yorkshire squid had not been published when Frost described the 1933 Newfoundland specimen, but two years later, when another giant squid washed ashore in Newfoundland, she was able to compare it with Robson's description of *Architeuthis clarkei.* The 1935 specimen "was seen floundering in the water at some distance from the shore at Harbour Main, Conception Bay. The animal must have been in the last stages of exhaustion, for Jos. Exekiel, a young fisherman, was able to drag it to the beach with the aid of a small boat" (Frost 1936). This was an immature female, with a mantle that measured 7 feet, 2 inches, and although the club end was missing, the longest tentacle was 17 feet long. Although Nancy Frost wrote, "The form of the fins differentiated it at once from that described previously from Dildo," she was unable to make a better identification. She concluded, "In many respects the species resembles *Architeuthis clarkei* Robson, but the radula differs considerably. In view of the lack of comparable data and because at present the species of the genus *Architeuthis* are not well defined, the specimen is described but no attempt is made to determine the species."

In July 1935, the fishing trawler *Palombe* out of La Rochelle snagged a *calmar géant* and brought the carcass back to the dock. Jean Cadenat, a scientist at the La Rochelle Laboratory of the Office of Marine Fisheries, described the capture:

> On July 1, 1935, the La Rochelle trawler *Palombe* (Capt. Le Bescou) of the fleet of F. Castaing, brought back to the company's warehouse a giant decapod cephalopod. It was more than eight meters long from the posterior end of its body to the tip of its extended long tentacles. It had been caught with an otter trawl, on June 8, 1935, in the Bay of Biscay, by 46°50' North latitude at a depth of about 200 meters. The weather was not particularly good, but nevertheless calm enough to allow normal trawling; this calm period followed a rather long stormy period.
>
> No one among the trawler's crew, not even the old sea hands who had fished on the banks of Newfoundland, where the presence of giant squids has been

mentioned, had ever found themselves in the presence of such a large squid. Unfortunately, the arms were cut up and the eyes completely torn off by the crew members. Moreover, before deciding to bring it back to La Rochelle, the body was left exposed to the air on deck for nearly forty eight hours, which accounts for its poor state of preservation when it reached us for examination.

Architeuthis even made an appearance in unexpectedly warmer waters. Gilbert L. Voss, one of the country's foremost teuthologists, wrote his doctoral dissertation on the cephalopods of the Gulf of Mexico. For his research, he examined some three hundred specimens collected between 1950 and 1956 by the U.S. Fish and Wildlife Service vessel *Oregon,* including a rather large one that was found floating on the surface of the delta of the Mississippi River. Although it was in poor condition—the head was nearly severed from the mantle and the arms and tentacles were floating free—there was no question about its identification: it was *Architeuthis.* (Voss identified it as *A. physeteris,* but recent taxonomic revisions suggest that this may not be a valid species.) The mantle was 2 feet (61.2 cm) in length, and the tentacles were about 9 feet long. Most of the skin of the mantle was present, and was observed to be reddish brown dorsally and yellowish, with minute reddish pigment spots, on the underside. The stomach was empty, but there were about thirty spermatophores projecting from the end of the penis. Each of these spermatophores was about 4 inches long. (See pages 110–111 for a discussion of the spermatophores of *Architeuthis.*)

In recent years, more giant squid have washed ashore or have been found floating at the surface. In 1955, a year after the 30-footer stranded at Ranheim, Norway, a 405-pound *Architeuthis* was recovered from the stomach of a sperm whale harpooned off the Azores. (If nothing else, this 405-pound item from the stomach of a cachalot demonstrates that a sperm whale is fully capable of swallowing a man.) At his laboratory at the University of Miami, Gilbert Voss was the recipient of a 47-footer that had been picked up floating on the surface northeast of the Bahamas in 1958, after what appeared to have been a battle with a sperm whale. In a little book called *Fantasi og virkelighed i naturen* (Fantasy and Reality in Nature), the Danish zoologist Lars Thomas wrote that "in Danish papers in 1965 one could read about a Jutlandic fishing skipper who had landed an 11 meter [36 foot] long squid in Skagen Harbor. It was caught in a net off

In early February 1980, a 30-foot-long giant squid carcass was found on the beach at Plum Island, Massachusetts. After being exhibited at the New England Aquarium in Boston, the 450-pound squid was transferred to the Smithsonian Institution in Washington, D.C.

Lindesnes in Norway. The skipper got 80 øre [about twelve cents] for each of the squid's 82 kilos [180 pounds], and the animal was thereafter sent to Italy, where sales possibilities were considered to be better than in Denmark." (It is conceivable—but not likely—that the Italians have a fondness for ammonium-flavored calamari.)

As predicted by Frederick Aldrich, fifteen specimens were recorded in Newfoundland between 1964 and 1982, some of which were in such poor condition that they could not be preserved.* In February 1980, a 12-foot specimen (its long feeding tentacles had broken off) with 10-inch-diameter eyes washed ashore at Plum Island on the Massachusetts coast. Scientists estimated that this

---

*One of these specimens, found floating at the surface in White Bay, Newfoundland, in 1964, had tentacular clubs that were so different in size and configuration that Frederick Aldrich and his wife, Margueritte, published a paper on the hitherto unsuspected ability of the giant squid to regenerate lost tentacles.

half-grown individual would have measured about 30 feet with its tentacles intact. (This is the specimen that is on exhibit at the Smithsonian.) And then in August 1982, fisherman Rune Ystebo looked out the window of his home near Bergen, Norway, and saw what he thought was a group of divers. He launched his little fishing boat to investigate, and when he realized that it was a live giant squid, he speared it. He dragged it to shore, where it expired; weighed it at 485 pounds; and measured it at 33 feet total length. Upon examining the specimen, Dr. Ole Brix, director of the zoological laboratory at the University of Bergen, ran a blood analysis and concluded that the oxygen capacity of the giant squid's blood was severely limited, which led him to claim that the animal is indeed a slow swimmer, and perhaps even a passive predator.

Martin Von Stanke, a South Australian commercial fisherman, encountered this specimen of Architeuthis *recently dead and floating just south of Mount Gambier in March 1995.*

Recently, climatologists M. E. Schlesinger and N. Ramankutty of the University of Illinois have identified forces that affect the temperature of the North Atlantic, which may somehow be related to the life cycle of the giant squid. As warmer water flows north from the equator—in the Gulf Stream, for example—it cools and sinks, pulling more warm water northward, like a conveyor belt. As the warm water builds up, it begins to rotate clockwise, drawing in new, lighter water from the west. This water cannot sink until it cools, so the process is slowed down until the newer water is cooled and turns counterclockwise, bringing in denser, saltier water from the east. (The eastern Atlantic is saltier than the western, because the salty, shallow Mediterranean flows into it.) Schlesinger and Ramankutty tentatively identified a seventy-year cycle for these changes, but physicists at the Geophysical Fluid Dynamics Laboratory at Princeton University created computer simulations of the cycle, and suggested that the cycle took anywhere from forty to sixty years (Delworth *et al.*, 1993). Nothing in these studies mentions giant squid, but the movement of the water in which *Architeuthis* lives must have some influence on the lives (and deaths) of these perpetually enigmatic creatures.

On March 9, 1995, about three miles off the town of Mount Gambier, South Australia, the carcass of a recently dead 30-foot female—identifiable by the presence of eggs—was found floating on the surface. It was collected by fisherman Martin Von Stanke, who brought it to the South Australian Museum in Adelaide, where Wolfgang Zeidler cataloged it for the collection. The head and body measured 6 feet in length, the longest tentacle was 24 feet long, and the eye was 6 inches across. The carcass was in good condition, except for two large gashes through the left side of the mantle, suggesting "that it might have been attacked by a predator such as a sperm whale" (Zeidler and Gowlett-Holmes, 1996).

When R. K. Dell (1952) summarized the *Architeuthis* situation in New Zealand waters, he relied on remarks that had been made by Anton Bruun about the cephalopods of Iceland. In his 1945 paper, Bruun had written, "From the distribution of the localities in which *Architeuthis* has drifted ashore, one might conclude that it were an arctic species; however, this is not the case, though very little is known about its natural habitat. Practically all records of *Architeuthis* . . . are indicative of a subtropical origin or at least an origin in fairly warm waters. The explanation of the numerous finds of *Architeuthis* on the North Coast of Iceland must then be that specimens from southern latitudes, perhaps run astray when hunting the herring, became paralysed when they came into contact with true arctic water." If Bruun was correct, wrote Dell, and *Architeuthis* lives normally in subtropical waters, then the giant squid in New Zealand ought to be concentrated in two areas: Cook Strait (between North and South Islands); and Foveaux Strait, between the southern tip of South Island and Stewart Island. He wrote: "There is another possibility—at least in regard to local occurrences. The two areas mentioned lie in close proximity to oceanic waters of great depth—e.g., more than

2,000 fathoms in the Cook Strait approaches. If the giant squids inhabit areas of great depth, they might be incapacitated when forced into areas of shallow seas."

In their 1994 paper, Gauldie, West, and Förch (discussed on page 6) reported on their examination of the statolith of a specimen of "*Architeuthis kirki.*" Aside from the questionable designation of *A. kirki,* this paper contained the startling news that "*a total of 24 specimens of giant squid were recovered between 1983 and 1988 in New Zealand waters*" (italics mine). The "zoological data will be published separately," so for the moment, we have only the capture date, location, sex, and dorsal mantle length of the two dozen giant squid. Of the 1983–88 strandings, five took place in the vicinity of Foveaux Strait, and only one on the shores of Cook Strait. The largest was a female that measured 83.46 inches (just under 7 feet), and the smallest was another female whose mantle measured 36.27 inches. According to the identified locations (most of the specimens were caught by trawl fishermen fishing for other things, but two were found washed up on shore), they can be found almost all around North and South Islands, and also in the vicinity of the Auckland Islands, some three hundred miles to the south.

New Zealand deep-water fishermen were hauling in giant squid at an unprecedented rate. The squid had apparently been feeding on schools of fishes (ling, orange roughy, and a hakelike fish known as the *hoki*) at depths of one thousand to four thousand feet. Now, for the first time, scientists had an idea of where these heretofore reclusive creatures might be found, and they were not content to wait for the fishermen to donate the serendipitously caught squid to their laboratories.

In January 1996, a large female *Architeuthis* was caught by Australian fishermen off the coast of Tasmania. Two more females were caught in Tasmanian waters that same year. Also in January 1996—the height of the antipodean summer—two more females, 26 and 13 feet long, were caught in the waters of the Chatham Rise, a rocky plateau east of New Zealand. Then a 20-foot-long male was hauled in from a trawl that had been set about a thousand feet down, also off New Zealand. A total of four giant squid was caught in New Zealand waters in 1996.

What does the increased presence of *Architeuthis* in antipodean waters mean? Has the home range of the giant squid shifted, or are Southern Hemisphere fishermen just luckier than their northern colleagues? Just as the 1870s saw a large number of giants on Newfoundland's rocky shores, we seem to be witnessing an ephemeral phenomenon, perhaps related to still unidentified ocean thermal patterns, or perhaps having something to do with breeding. Architeuthids have appeared in New Zealand waters before; from 1870 to 1887, there were seven records, including the 1886 specimen recorded by C. W. Robson and named for T. W. Kirk. More giant squid have been recorded from Australian and New Zealand waters from 1983 to 1996 than anywhere else in the world for a similar time period. (In Newfoundland waters from 1870 to 1881, the confirmed total was twenty, but others were not recorded by scientists or were cut up for bait.) Once again, *Architeuthis* has confounded us with its unwillingness to conform to any recognizable pattern; it refuses to play by the rules.

In January 1997, an expedition was mounted to attempt to film *Architeuthis* in what was supposed to be its natural habitat. Led by Clyde Roper, the expedition consisted of Malcolm Clarke, Greg Stone of the New England Aquarium, National Geographic photographer Emory Kristof, Teddy Tucker of Bermuda, Adam Frankel of Cornell University's Bioacoustic Research Program, James Bellingham of MIT's Underwater Vehicles Laboratory, and various support teams, including programmers who were to send back daily dispatches to Smithsonian and National Geographic Web sites. They were going to New Zealand; Kaikoura, on the northeastern coast of South Island, was to be their jumping-off point, since it is on the edge of the Kaikoura Canyon, a deep trench known to be the year-round haunt of sperm whales. There they planned to approach the problem from several angles. MIT's *Odyssey,* an autonomous underwater vehicle (AUV), would be deployed with a video camera attached, in hope of filming a squid. (The AUV can also measure water temperature and salinity to get an accurate picture of the type of habitat the squid prefers.) Emory Kristof planned to drop several baited cameras. Each camera looked, he said, "like a 4½-foot box kite flying horizontally in the water" and contained a six-gallon drum of smelly, liquefied fish, as bait for the squid. And finally, they were going to try to affix a video camera (the "crittercam") to a sperm whale, and as the whale descended in search of food it was going to film its prey. The camera is free-floating, attached harmlessly by a tag inserted into the whale's back by a short length of line, and designed to face the same direction as its host. The camera has a link that dissolves in seawater after two hours, and a radio beacon, which enables the researchers to collect it when it has come loose and floated to the surface, so they can retrieve the tape.*

In Gauldie, West, and Förch's 1994 discussion of the statocysts and statoliths of giant squid, there is a map that shows the locations of the twenty-four specimens that were collected between 1983 and 1988. Of these, four were found in the deep waters off Kaikoura. It seemed a combination almost guaranteed to succeed: the known habitat of sperm whales and giant squid, experts who had studied extensively the habits of these two great monsters of the deep, and high-tech equipment to get the first-ever images.

While the whales were always visible, no giant squid appeared, and by the time the New Zealand Department of Conservation granted permission to use the crittercam, the weather had so deteriorated that the whales could not be safely approached.

---

*Although this sounds like a mad scientist scheme, it was actually successfully employed with great white sharks in South African waters, for the spectacular 1995 National Geographic film *Great White Shark,* and again with sperm whales in the Azores for the 1998 *Sea Monsters: The Search for the Giant Squid.*

What Do We Know About Architeuthis?

All we really know about the natural history of the giant squid is that it occasionally washes ashore—and when that happens, we don't know why. Its feeding habits, breeding habits, vertical and geographic distribution, life span, and habitat are all unknown. Furthermore, there are some areas where the nature of the animal itself almost precludes our obtaining any but the most rudimentary information. Take size, for example. While the mantle is a tough, muscular structure that cannot be easily deformed, the tentacles can be stretched like gigantic rubber bands, allowing for great exaggerations of the animal's total length. Therefore, mantle length (abbreviated "ML") is usually considered the most reliable length measurement, although tentacle length, which varies greatly according to species, is also important in identification.

As with so many aspects of giant squid lore, the squid's weight is a subject of vast ambiguity. In Nettie and G. E. MacGintie's 1949 *Natural History of Marine Animals,* for example, we read that "two arms of *Architeuthis* that were 42 feet long were found, and if one reconstructed a body . . . the squid to which these arms belonged was 4.6 feet in diameter and 24 feet long, with an overall measurement of 66 feet. It would have weighed about 42½ tons." In the MacGinties' book we read that a 55-footer "would have weighed 29¼ or 30 tons [a curious and unexplained range] including the tentacles—a truly noble animal, being a little more than one-fifth the weight of the largest whale and larger than the whale sharks and basking sharks, the largest of all fishes." These figures seem to be unfounded exaggerations. Where squid carcasses have actually been weighed, it appears that the longest ones—in the 50-foot range, for example—weigh about a ton.

Bernard Heuvelmans, however, begs to differ. His 1958 book *Le kraken et la poulpe colossal* (translated into English in 1997) is filled with accounts of giant squid and parts thereof that strongly suggest to him that "there must be *Architeuthis* weighing more than 5 tons, and some even larger ones which must weigh between 2 and 27 tons, the normal weight being around 8 tons. *There are good reasons to believe that there may even exist specimens twice as long as that of*

Thimble Tickle, which, depending upon their girth, might have weighed between 16 and 216 tons, but more likely around 64 tons." I have added the italics to Heuvelmans's assertion because I think it is utterly ridiculous. It is impossible to demonstrate that a 16-ton *Architeuthis* does not exist, but even if it did, Heuvelmans commits a fundamental error in calculating the weight of some of these monsters when he writes that "the density of living creatures is only slightly higher than that of water . . . a decimetre of living flesh weighs about as much as a litre of water." That may be true for some other living creatures, but the flesh of *Architeuthis,* saturated with ammonium chloride, is *lighter* than water, and the giant squid is neutrally buoyant. (This is believed to be the reason that dead or dying squid are found floating at the surface or washed up on the beach.) His assumption, therefore, that the 55-foot-long Thimble Tickle squid would have "probably weighed near 24 tons" is patently erroneous.

Born in 1916, Heuvelmans is perhaps the world's foremost cryptozoologist— in fact, he invented the term and the discipline, which he defined as "the study of hidden or unknown animals." His books* are paradigms of careful research and meticulous documentation. When he established the Center for Cryptozoology at Le Vésinet, outside Paris, he brought with him "about 2,500 volumes, of which some 800 were in the field of cryptozoology, some 100,000 original or photo-copied pages on cryptozoology; some 2,500 bibliographic filing cards, and an incomparable documentation of about 4,000 illustrations and photos."

Some zoologists, crypto- and otherwise, believe only in those creatures (or their dimensions) that can be empirically verified. Thus, the coelacanth and the megamouth shark, earlier believed to be extinct or nonexistent, respectively, were shown to be living creatures that could be examined, measured, photographed, and stored in museum collections, and are now universally recognized as legitimate members of the world's large marine fauna. Similarly, despite stories about 30- or 40-foot-long great white sharks, the accepted record size for *Carcharodon*

---

*Heuvelmans's books, which are a veritable encyclopedia of cryptozoology, have been published and republished in various combinations, so a definitive list is difficult to prepare. According to his own catalog, included in the introduction to a 1995 republication of *On the Track of Unknown Animals* (originally published in 1955), his books are: *The Kraken and the Colossal Octopus* (1958); *In the Wake of the Sea-Serpents* (1965); *The Last Dragons of Africa* (1978); and *Beast-Men and Man-Beasts of Africa* (1980).

carcharias is 21 feet. But Heuvelmans, who has written numerous books and articles on mysterious animals, is of the school that maintains that anything can exist—at almost any size—because no one can prove it doesn't. His position on *giant* giant squid—and it is a hard one to argue with—is that those reports of huge arm fragments, sightings at sea, and even attacked ships cannot all be hoaxes, misidentifications, or typographical errors. (The same goes for sea serpents, gigantic hominids traipsing through the snowfields of the Himalayas, lake monsters in Scotland, dinosaurs in the African jungles, and any number of other cryptozoological entities.) As evidence that cryptozoology is a valid pursuit, he cites not only the coelacanth and megamouth, but the hundreds of unexpected creatures that have been discovered, such as the okapi, several new species of beaked whale, two new lemurs in Madagascar, and the two new species of deer that were found in the jungles of Vietnam. Of course, the presence of the okapi in the African forests does not prove that a dinosaur lives in more or less the same place, but it does suggest that there are many creatures still to be discovered in the more remote and unexplored parts of the world. I'm sure there are, but that doesn't mean one of them will be a 216-ton squid.

It is certainly helpful in identifying a particular animal if we know what color it is supposed to be. Although we must infer this from our limited knowledge of other large teuthids, we assume that *Architeuthis* can change color more or less at will, so there can be no definite identification based on color. Most of the dead or dying specimens showed some traces of reddish coloration, and in their 1982 article in *Scientific American,* Roper and Boss wrote that "the multilayered integument that envelops the body, the head and the arms is a dark purplish red to maroon dorsally and slightly lighter ventrally." In her 1948 description of a giant squid that was found at Victoria, Australia, Joyce Allan wrote of the animal's coloration: "The living animal must be simply amazing to witness. Remains of the outer skin (underneath this the flesh was firm, smooth, and blanc-mange white) was brilliant carmine red, due to a minute speckling of that colour." When fishermen in Trinity Bay, Newfoundland, encountered a live giant squid at the surface, they saw that it displayed "vivid color changes" before it dived and disappeared. When Joakim Lerøen gaffed a squid that was swimming back and forth in Norwegian waters, it was light-colored, "but when the gaff cut into its eye, it turned deep purple" (see pages 94–95).

Since *Architeuthis* can change color, descriptions of its basic color are bound to vary from one account to the next. Also, we do not know if a certain color occurs when it is injured or dying, and since most known specimens have been dead or moribund, our information may be skewed. In any event, most descriptions agree with Roper and Boss and put it in the red to maroon range:

> The multilayered integument that envelops the head and the arms is a dark purplish red to maroon dorsally and slightly lighter ventrally. The color of the dorsal and ventral surface of the arms is less intense than that of the lateral ones. . . . The internal surface of the mantle and some of the viscera also have dark reddish pigmentation, an uncommon feature in oceanic squids.

The stomachs of most stranded giant squid have been empty, so we have little idea of what they eat, but from fishing experiences, it has been observed that other species, such as the 10-foot-long Humboldt squid *(Dosidicus gigas)* are cannibalistic. Because of the radular "teeth" on the tongue and on the pharynx, chunks of food bitten off by the powerful beaks are shredded before entering the alimentary canal. This is one of the reasons why the food items of squid are so difficult to identify.

Deep-water animals, whether fishes or cephalopods, usually have larger eyes than their shallow-water counterparts. Since *Architeuthis* has the largest eyes of any animal on earth, it follows that it must use these eyes to see where there is limited light. It does not follow, however, that the eyes are large simply because the squid is large. The sperm whale, which can reach approximately the same length as the giant squid and, at least on some occasions, frequents the same depths, has an eye about 2½ inches in diameter, not much larger than the eye of a cow. The eyes of a giant squid can be as much as 15 inches in diameter, larger than an automobile hubcap. The eye of a blue whale, the largest *animal* in history, gets to be about 7 inches across.

Many species of squid can bioluminesce by lighting up special cells known as photophores, but *Architeuthis* is not numbered among them. While the giant squid does have an ink sac, it is quite small and suggests that the animal does not eject clouds of ink that are commensurate with its size. In the lightless depths of the ocean the giant squid has no known predators except the sperm whale, and an animal that hunts by sound in total darkness is not going to be deterred by a cloud of ink or a change of color.

A number of squid species are not neutrally buoyant and have to keep moving just to remain in place. If they do not maintain constant motion, usually by a combination of fin movement and ejections from the siphon, they sink to the bottom. This is not the case for *Architeuthis*.

The giant squid is rubbery and heavy out of the water, but because its muscles are filled with vacuoles containing a solution of ammonium chloride, which is lighter than water, it has a neutral buoyancy in water, and will float upward if it does not make an effort to stay down. Many squids are ammoniacal, and according to a 1979 study conducted by Clarke, Denton, and Gilpin-Brown, "no less than 12 of the 26 families achieve near-neutral buoyancy," while the other 14 are denser than seawater and must swim to keep from sinking. The ammoniacal squids, of which *Architeuthis* is but one, are, according to the Russian teuthologist Kir Nesis, "not edible to humans, though they might satisfy the taste of the sperm whale."* Neutrally buoyant species of the families Histioteuthidae, Octopoteuthidae, and Cranchiidae are important in the diet of sperm whales, wrote Clarke *et al.,* "and the combined weight of these three families consumed each year by these whales probably exceeds the total fish catch of all the fishing fleets of the world." Is there a connection between the ammonia and the sperm whale's preference, or do neutrally buoyant squid, suspended in the water column, present a better target for the sonic blasts of hunting sperm whales? The ammonia in the tissues of the squids tested by Clarke, Denton, and Gilpin-Brown was found when the squids were cut and their body fluids analyzed. The more liquid was removed from a particular tissue, the less buoyant it became.

The examination of stranded specimens of *Architeuthis* has provided some of the requisite anatomical data, but otherwise very little is known about the reproduction of giant squid. Most of our information—with the exception of the

---

*Upon the completion of his doctoral dissertation (on the Atlantic squid genus *Illex*) in 1973, C. C. Lu celebrated by cooking a piece of *Architeuthis* for himself, his wife, and the other teuthologists on his committee, Clyde Roper, Malcolm Clarke, and Fred Aldrich. It proved to be inedible because of the "strong, bitter taste of ammonia," but Clarke (with Denton and Gilpin-Brown) brought it back to England to analyze it, and eventually published their paper on the use of ammonium for buoyancy in squids.

description of the physical elements—has been taken from accounts of other, more accessible species, and should therefore be considered conjectural for *Architeuthis*. Mature male squids (and octopuses) have one or more of their arms modified to form a hectocotylus,* a sex organ used to transfer spermatophores (long tubes filled with sperm) to the female. Males, which can be easily differentiated from females by these two arms with flattened tips and no suckers, store the spermatophores in an organ known as Needham's sac (because it was discovered by one John Needham). Then—in most species—the male uses his specialized arm to transfer the spermatophores to the oviduct of the female. The spermatophores of *Architeuthis* are tightly coiled and between 4 and 8 inches long, which seems more than large enough, but the Pacific octopus *(Octopus dofleini)* has spermatophores that may measure nearly 4 *feet* in length when unrolled.

Architeuthis is probably a solitary hunter, but there has to be a time, however fleeting, when two of them get together. As with most aspects of the biology of the giant squid, we can only speculate on what transpires at those times, but there are some smaller species of squid whose mating activities have been closely observed. Off the coast of southern California in the spring, the market squid, *Loligo opalescens*, gathers for its annual spawning and mating rituals. In recent years, diving into and filming the mating frenzy of the squid has become a popular activity, so there is no shortage of images, still and moving, of squid *in flagrante delicto*. They congregate in enormous numbers; "We were in the middle of an almost solid layer," wrote Cousteau and Diolé, describing their dive, "several yards thick, of writhing squirming creatures who darted to and fro by expelling water from their funnels, like a vast fleet of miniature jets."

With his arms facing those of his partner, the male *Loligo* holds the female,

---

*The word *hectocotylus* has a fascinating history. In 1829, Baron Cuvier examined several argonauts (highly specialized octopuses also known as paper nautiluses because of their fragile shells), and found what he believed to be a parasitic worm. Because the worm *resembled* the arm of a cephalopod, he called it *Hectocotylus*, which means "a hundred suckers" (*hecto*, "hundred"; *cotyla*, "cup"). It was not until 1853 that the German zoologist Heinrich Müller recognized that this "worm" was the sex organ of the tiny male argonaut. Displaying the greatest sexual dimorphism of any cephalopod, the male is only an inch long, compared to the female's eighteen-inch bulk, including tentacles. His reproductive organ—known as the *hectocotylus*—breaks off and is carried around in the shell of the female so that fertilization can occur. Thus, *hectocotylus*, and its adverbial form, *hectocotylized*, have come to describe the arms specially adapted for the spermatophore transfer of many male cephalopods.

and using the tip of his hectocotylized arm, he picks up a batch of spermatophores and inserts them into the female's mantle, close to the funnel. Fertilization of the eggs is the trigger for their being laid, and the females attach the egg cases to the ocean floor with an adhesive stalk. The Cousteau team described the clusters of egg cases as "looking like dahlias." After the copulation and egg laying, the squid are substantially weakened, and those that do not die outright are easy prey for the sea lions, sharks, and dolphins that hungrily attend the annual spawning. There is something about the egg cases that makes them completely unappetizing to predators, so they remain untouched on the bottom for three to five weeks, until the eggs hatch, and the growth and reproduction cycle begins again.

Giant squid most likely perform a more cumbersome, less populous version of the mating death dance of *Loligo*, but we must strain our imaginations to conjure up a picture of two of these apparitions (the females of most squid species

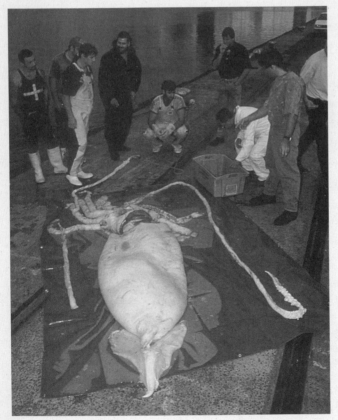

Scientists examine one of the female giant squid captured off Tasmania in 1996. Most of the skin was missing from this specimen, which accounts for its almost white coloration.

are larger than the males) locked in a twenty-arm embrace, where the tail fins of the female can be sixty feet from those of her mate. Do they flush an unseen crimson in the blackness of the abyss? Do they, like many other species, die after mating? Are some of the dead or dying giants that have washed ashore casualties of this titanic battle of the sexes?

From circumstantial evidence, however, researchers are getting an idea of what happens when giant squid engage in sex, and it is weirder than anyone could have imagined. As reported in a 1997 article in *Nature,* Mark Norman and C. C. Lu (both of Melbourne, but Lu has now moved to Taiwan) examined two large females that had been captured in between sixteen hundred and twenty-five hundred feet of water off southern Australia, and one of them had spermatophores embedded in the skin of both ventral arms. The location of the spermatophores on the ventral arms—and nowhere else—suggested that they had been placed there deliberately by the male. (The females of other squid species are known to store spermatophores on the mantle, the head, the arms, around the mouth, within the mantle cavity, or within the reproductive tract, and from limited observations, it appears that *Architeuthis* does the same.) The penis of a mature male giant squid can be 3 feet long, and even though they have also been observed with hectocotylized arms, they may use the penis to transfer spermatophores directly to the female. Without any sort of hook to create an opening into which the spermatophores might be placed, it is possible that the male *Architeuthis* "injects" the sperm, "potentially under hydraulic pressure," into the skin of the female, and she somehow (this part is very unclear) takes it out and places it in her reproductive tract to fertilize the eggs. Norman and Lu concluded their article:

> As spawning events have never been witnessed for "skin-storing" squids (including *Architeuthis*), it is not known how females access sperm to fertilize their eggs. The suckers or beak may be used to peel open the skin covering the spermatophores, or the sperm may migrate to the surface on hormonal or chemical cues. Alternatively, the female's skin might degrade on spawning, exposing the embedded sperm stores.

The female squid discussed by Norman and Lu was immature, but her ovary contained hundreds of thousands of undeveloped eggs. "The discovery of sperm," wrote the authors, "suggests that female *Architeuthis* can store sperm from sub-mature stages, as reported in other cephalopods, perhaps reflecting that encounters with male giant squids are infrequent at these dark depths."*

A large female washed ashore at Cove Bay, south of Aberdeen, Scotland, on January 8, 1984, and was described by P. R. Boyle of the University of Aberdeen. It was 4.23 meters (13.87 feet) overall, with a mantle length of 1.75 meters (5.74 feet). Both tentacles were missing. Its total weight was 168.4 kilograms (371 pounds), and because of freezing temperatures and snowstorms, it proved to be difficult to lift out of the water. It was a female, with approximately three thousand eggs attached to three detached egg strands. Where Ole Brix had suggested that giant squid suffocate near the surface because the hemocyanin content of the blood requires the cold water temperatures associated with the depths, Boyle wrote that "the low water temperature associated with the Scottish stranding does not support the proposal that mortality of *Architeuthis* at the surface is due to asphyxia caused by the low temperature of oxygen affinity of the hemocyanin."

A baby giant squid is almost a contradiction in terms; we have been so conditioned to think of these creatures as formidable monsters that it is almost impossible to envision a baby. But in 1981, the Australian research vessel *Soela* pulled in its nets and found, among other things, a tiny squid with a mantle length of 10.3 mm—just under half an inch. In the minds of the Australian scientists, a half-inch squid had nothing whatever to do with the giant *Architeuthis*, but when C. C. Lu examined it closely "during a routine sorting of cephalopod specimens lodged in the Museum of Victoria," he realized it was actually a larval *Architeuthis*—the smallest one ever seen. And it was netted at a depth of twenty meters, "the shallowest known capture depth of any *Architeuthis*."

"The specimen differs from large specimens," wrote Lu, "in several anatomical features and body proportions. The characteristic carpal cluster of suckers and knobs on the tentacle are not found. On the tentacular stalk, the paired suckers

---

*When the *Times* of London covered this story, Science Editor Nigel Hawkes entitled his story "Sex Is a Shot in the Arm for Giants of the Ocean." Something about the story must bring out the humorist in writers, for even Norman and Lu, discussing a male squid that had been found in Norway with spermatophores embedded in its arms, wrote, "Another male may have injected the spermatophores while attempting to impregnate a female, accidentally 'riveting' a co-suitor. Alternatively, the male may have literally 'shot himself in the foot.'"

and knobs found in adults are not developed, although the precursors of suckers are discernible. In the adults the fins are longer than wide, while in the present specimen the fin width is about 2.5 times the fin length."*

Two juvenile giant squids have been examined, both now in the collection of the Institute of Marine Sciences at the University of Miami. (Neither was *collected* by Miami scientists; both were found in the stomach contents of fishes in the ichthyological collections.) The larger of the two was 57 mm (2.2 inches) in mantle length, and had been taken from the stomach of a longnose lancetfish *(Alepisaurus ferox)* captured off Madeira. The second specimen was taken from the stomach of a fish—also probably *Alepisaurus†*—in the eastern Pacific off Chile, and was 45 mm (1.75 inches) in mantle length. Until Lu's larval specimen appeared in 1986, these were the smallest ever measured. As Roper and Young put it in their 1972 report, "they are an order of magnitude smaller than the smallest previously reported specimen," a 460 mm (1.5-foot) specimen of *A. physeteris.*

We know so little about the biology of *Architeuthis* that the subject is ripe for speculation. Anyone who has ever examined a dead specimen (and that means anyone who has ever seen a giant squid in a lab or on the beach) has theories about the lifestyle of these mysterious creatures. Are they aggressive hunters? What do they eat? How long do they live? Where do they breed?

The four giant squid caught in nets off New Zealand in the early months of 1996 were studied by Steve O'Shea of New Zealand's National Institute of Water and Atmospheric Research. None of the squid was caught intentionally; all were

---

*How did he know it was *Architeuthis?* There are several diagnostic characteristics of the species, delineated in a later paper by Jackson, Lu, and Dunning: "A straight simple funnel-locking cartilage, the buccal connectives were attached at the dorsal border of arm IV, and the tentacular club had four rows of suckers, with those on the medial rows of the manus much larger and those on the marginal rows small. Also, a distinct cluster of numerous small suckers and knobs were at the proximal end of the manus, and two longitudinal rows of alternating suckers and pads were on the tentacular clubs."

†What sort of a creature eats giant squid? The lancetfish is described as "a fierce looking mesopelagic carnivore that is one of the largest species found in the midwater fauna, averaging 1 to 1.5m in length and occasionally exceeding 2m. . . . Lancetfishes are voracious carnivores, actively pursuing all smaller fishes and often eating smaller individuals of their own kind" (Ayling 1982). Frank Lane (1974) quotes teuthologist Gilbert Voss as saying, "Stripping the stomachs of these fish is one of the approved methods of collecting deep-sea squid and octopods."

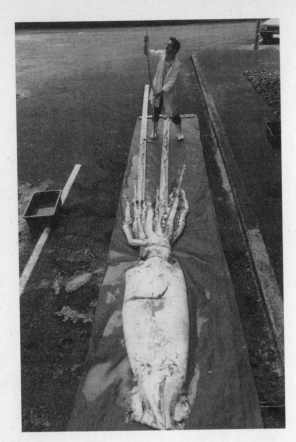

Squid specialist Steve O'Shea holding up the end of one of the tentacles of a 26-foot-long giant squid captured off New Zealand in 1996.

acquired in commercial fishing operations. The only food items found in the stomachs were a few fish scales, which led O'Shea to write, "The stomach was full of a fluid which did not look at all like any dietary composition. This is seen time and time again in other *Architeuthis* stomach contents, almost always empty."*

One of the New Zealand specimens was a 20-foot male (caught on January 16, 1996) that O'Shea says was

caught with his penis hanging straight out through the funnel. Spermatophores were everywhere over the carcass when I got it. [See pages 113–114.] This is interesting as I believe a humongous female was right next to

---

*The information from Steve O'Shea was included in various E-mail messages to me, in May and June 1996, in response to my request for information on these New Zealand specimens.

him in the water column when the nets went by. He was probably so orgasmically stupefied that he just didn't see the net coming.

He further speculates:

> I believe these animals probably live deep in the abyss and come onto the continental shelf to breed. This may explain why the stomachs are always empty—they are not feeding but breeding, depending on their huge size to keep them going. . . . Given the more than coincidental capture of a number of specimens off South Australia, we must be dealing with a single species with enormous distribution—probably abyssal, probably oceanic—for at least a part of its life cycle. The eggs are extremely tiny, 1–2 mm in maximum dimension, so this would also indicate a prolonged larval stage. The most recent research on the animals indicates a short life span (ca. 3 years); they do not have the opportunity to live to 20 before bumping into a prospective mate in the dark abyss. I am not saying these things are common and school, but I believe they are far more common than we presently realize.

The single larval specimen examined by C. C. Lu shows that *Architeuthis* is tiny at birth. We know nothing else about the spawning process, but in their discussion of the two juveniles, Roper and Young suggest that "the small size of the present specimens suggests they were spawned in the nearby region of their localities of capture in tropical or warm temperate waters." The lancetfishes that swallowed these two juveniles are known to hunt primarily in the upper three hundred meters (one thousand feet) of the open ocean, so this gives us a fleeting clue about the vertical distribution of these otherwise enigmatic decapods. As they grow, they may migrate into deeper waters. Lu wrote: "The day time capture [of the larval specimen] at such a shallow depth suggests that *Architeuthis* may exhibit ontogenetic descent and the larvae may be near surface dwellers."

The presence of *Architeuthis* beaks in the stomachs of albatrosses examined by M. J. Imber also supports the idea that the younger ones live closer to the surface. In his 1992 analysis of the cephalopods eaten by wandering albatrosses *(Diomedea exulans)* at six circumpolar localities, Imber found that juvenile *Architeuthis* "were most important numerically, and also by biomass, in the diet at Gough and Macquarie Islands." (The other localities sampled were the Auckland Islands, the Antipodes, Prince Edward Island, and South Georgia.)

In 1881, Steenstrup reported on certain squids that had been collected by "First Mate Asm. Corneliussen . . . during his later voyages in the southern seas." These consisted of the remains of cephalopods (genus *Gonatus*) "which he had cut out of the stomachs of albatrosses, *Diomedea exulans*. These large birds belong, as is well known, to the most voracious cephalopod hunters, as their food consists mainly of the oceanic forms living far from the coasts. As a matter of course the content of their stomachs is of the greatest importance to the scientist whose knowledge of cephalopod forms from the open sea has hitherto been very scarce."

Occasionally, these large birds (the maximum wingspan for *D. exulans* is 11 feet, making it the largest of all flying birds) were seen to feed while sitting on the water, but they would also dive to pick up a morsel while on the wing. Nowadays, instead of cutting open the bird to see what it ate, researchers can examine the regurgitations of adults returning from foraging forays to feed their chicks on the nest, or induce the chicks to void their stomach contents. Of the squid species identified by the beaks, most are known to migrate toward the surface at night. Imber (1992) wrote that "upward movement at night takes some of these [squid species] into surface waters, including *Octopoteuthis* and *Taningia*. *Magistoteuthis* also move upwards, but with most below 200m at night, few might be available to albatrosses. *Gonatus antarcticus,* most of which live in the upper 250m, should reach the surface at night. Young *Architeuthis,* living at similar depth, may do the same."*

In the same communiqué discussed earlier, Steve O'Shea wrote:

> A colleague in a different NZ institute has approximately 20 *Architeuthis* beaks gathered from bird regurgitations off the Chatham Islands. If 20 juvenile specimens can be eaten by one bird yet we capture only four giants in a four-

---

*Neither albatrosses nor sperm whales appear to be allergic to ammonia in their diet, but it is not high on the list of preferred seasonings for human food. The squid species that people eat are nonammoniacal (which means they are heavier than seawater and have to keep swimming to prevent themselves from sinking), and include those taken by commercial squid fishers, such as the Loliginidae (the small squids usually served as calamari) and the larger *Ommastrephes, Todarodes,* and *Dosidicus.*

month period, it comes as no surprise that the birds are doing something different to capture these animals. If they were moribund at the surface (which they most likely were) these 20 beaks simply indicate that 20 dead individuals were caught at the surface. If 20 are dead, there must be hundreds below the surface. If population structure is anything to go by, I would expect many more juveniles than adults and more dead (spent) adults than juveniles.

O'Shea is the first to admit that these speculations are pure guesswork. (He says, "I'm sitting in an office in N.Z. so far removed from all else that I have a lot of time for the imagination to run wild.") He might be wrong; giant squid might not breed over the Chatham Rise, they might not quit feeding during the breeding season, or they might not live for only three years; but as an authority on *Architeuthis* (and one who has recently had four carcasses delivered to him), O'Shea has to be allowed flights of fancy within the framework of his own experience. Who, after all, is better qualified to speculate on the life and loves of *Architeuthis*? Jules Verne? Peter Benchley?

Squids specialize in a high reproductive rate and a short life cycle, as contrasted with fishes, which typically have long-lived adults that invest only as much in reproduction as they can afford to while ensuring their own survival. (In most of the squid species whose reproductive biology has been studied, death follows breeding.) Although there is no hard data to support such an assumption, some teuthologists believe that *Architeuthis* grows quickly and lives only a short time. If the giant squid is not an apex predator, it must pass quickly through the stages where it could be easily consumed by other predators. Until the relatively recent identification of larval and juvenile specimens, no one knew how small a newly hatched *Architeuthis* really was. Clyde Roper, who believes that the giant squid is not a powerful swimmer, is quoted as saying, "The more rapidly you can grow, the quicker you can reduce the number and kinds of other animals that can prey on you. With its growth rate, by the time it reaches adolescence, it is already too large to be eaten by anything other than toothed whales" (Abrahamson 1992).

Peter Boyle of the University of Aberdeen examined the carcass of a female giant squid that washed ashore at Cove Bay, on the east coast of northern Scotland in January 1984. Two of the arms were complete (the longer measured 7.5 feet in length), and the mantle was almost 6 feet long. Boyle found that the ovary was "packed with very small eggs, all 1 mm in diameter." The tiny eggs suggest that *Architeuthis* is born small and grows very rapidly. And indeed, when Jackson, Lu, and Dunning counted the growth rings on the statolith of a juvenile female (mantle length: 16.5 inches) that had been trawled off southern Australia in 1982, they concluded that the squid was 153 days old, and that it grew at a rate of one-tenth of an inch per day. (Counting the rings on the statolith had been demonstrated in other species to be a reliable indicator of growth.)

In "Statocyst, Statolith, and Age Estimation of the Giant Squid, *Architeuthis kirki*," Gauldie, West, and Förch discussed how they sectioned a statolith of one of the New Zealand specimens (an immature female) to determine the growth rate. Based on observations of captive animals of other species, they concluded that "young squid are capable of dramatically fast, exponential growth when food is not limiting." If the South Australian squid caught on January 30, 1992, had 153 microincrements on the statolith, and if these layers are deposited daily, then, Jackson *et al.* say, "the animal would have a putative birth date of 30 August 1991. . . . A review of whether microincrements are deposited daily in several species of squid can be found in Rodhouse and Hatfield (1990). The statolith described in this paper, which was collected off New Zealand on 3 May 1987, had 393 macroincrements. This would give a putative birth date of 6 April 1986." Gauldie, West, and Förch's New Zealand specimen had a mantle length of approximately 5 feet, which would mean that it grew from whatever size it was born at (an unknown) to the size it died at in one year and twenty-eight days. These three authors wrote, "The tissues of the squid mantle are highly collagenous and what little skeleton is present is in the form of cartilage. Both cartilage and collagen have very low energy demands (less than 1 percent of that of bone), and the consequent energy conversion efficiency of squid may be high. Very rapid growth would explain why so few small specimens of *Architeuthis* sp. have been reported."

Often overlooked in discussions of monster squid are occasional mentions of "miniature" or "pygmy" *Architeuthis*. In 1978 a miniature adult, with a mantle length of 6.5 inches, was found in the stomach of a 6.5-foot swordfish caught in Florida waters. According to Toll and Hess, who described it, "This specimen . . . is the smallest functionally mature *Architeuthis* known. Its size and state of maturity raise several questions regarding the life history of the 'giant squid.'" In his overview of the cephalopods of the world, the Russian teuthologist Kir Nesis wrote that "the pygmy giant squid may be an undescribed species." (It is certainly one of the English language's premier oxymorons.)*

Architeuthis has hardly ever been seen at the surface. Scientists are at a loss to explain this, but some have suggested that the squid can exist only at depth because they require the pressure to function, or that they thrive in the cold of the abyssal depths, and the warmer, deoxygenated waters nearer the surface may debilitate or even kill them. It is also possible that despite its great size and fearsome reputation, the giant squid is a relatively weak, slow-swimming animal that feeds at great depths and therefore has no reason to approach the surface unless it is sick or dying and cannot remain below.

Because the giant squid is still inadequately known, most of the discussions of this creature have been either fantasies about its attacking ships and dragging people from the shrouds, or detailed scientific descriptions of the carcasses that have washed ashore. There have been a few attempts to speculate on its abundance (assuming there is a single species), its habits, or its habitat. Especially noteworthy in his efforts to develop a more complete picture of *Architeuthis* was G. C. Robson of the British Natural History Museum, who was probably the first to try to extrapolate a "way of life" for the animal, based on his observations of its structure. His 1933 discussion included a section he called "General Characteristics and Presumed Mode of Life of the Group," in which he noted that the fins are small in relation to the total size of the body; the locking apparatus of the funnel mechanism is weak; the suckers are small and feeble; and the lateral and dorsal membranes of the arms are poorly developed when compared to other species. These characteristics led him to conclude that the giant squid was ill equipped to hunt large prey and was probably an inactive scavenger, waiting close to the bottom for invertebrates and carrion to come its way. He wrote:

---

*In 1952, Japanese scientists examined two specimens that had been found in the digestive canal of sperm whales caught off the Bonin Islands. The larger of the two measured 8 feet in total length, mantle *and* tentacles, and the second specimen was smaller. Eiji Iwai identified the specimens as "oegopsiden squid belonging to the genus *Architeuthis*." In a subsequent discussion, however, Roper and Young (1972) wrote that there is "no doubt that the identificaton is incorrect," and suggested that "the specimens appear to be members of the Psychroteuthidae, a little known family of oceanic squid, previously known only from Antarctic waters."

I am inclined on the whole to think that *Architeuthis* is rather a sluggish animal, living near the upper stretches of the continental slope in water between 100–200 fathoms, or deeper where the water temperature is high. The structure of the suckers suggests that it does not deal with large prey. The remarkably small size of the fins suggests an inactive life, so that it may keep near the bottom and feed on sendentary invertebrates and carrion.

Other teuthologists have also conjectured on its qualities and capabilities, with opinion seemingly divided into two camps. One that holds that *Architeuthis* is a powerful, aggressive predator; the other, that it is a weak, ineffectual hunter. Japetus Steenstrup believed that *Architeuthis* lived on the bottom in dark, deep water. In a 1959 magazine article, Gilbert Voss wrote, "In the giant squids . . . these organs [the mantle-locking cartilages, funnel valve, and so on] are very poorly developed; the funnel is flabby, the valve is weak, and the locking cartilages are mere shallow grooves and ridges."

Responding to these attacks on the poor squid's reputation, Malcolm Clarke (1966) wrote, "Stomachs [of stranded specimens] are almost invariably empty, but the suggestion that squid are probably poor swimmers and ill adapted for catching active prey (Robson, 1933; Voss, 1956) seems incompatible with the many hundred large suckers up to 3 cm diameter, the very powerful buccal muscles, the short, thick jaws giving maximum leverage, and the thick mantle wall."

In a study entitled "The Energetic Limits on Squid Distributions," Ron O'Dor wrote that

> *Architeuthis* could go around the world in 80 days (under the North Pole?).
> Why it would want to is unclear, but traveling from the northern bloom to the
> southern bloom [of plankton] is energetically feasible for some whales and could
> take *Architeuthis* less than a month. Such a pattern is consistent with its distrib-
> ution. It is now popular to say that *Architeuthis* is not a strong swimmer, but the
> evidence is only that it is ammoniacal and not as muscular, relatively, as the
> smaller squids. . . . At cruising speeds, squids use only 10% of the power (and
> presumably 10% of the muscle) available for an escape jet. Perhaps *Architeuthis*
> needs only cruising muscle, since there cannot be too many things it needs to
> escape from.

In his comprehensive 1991 discussion of *Architeuthis* in Newfoundland, Aldrich reviewed the earlier hypotheses about the behavior of the giant squid and wrote:

> My thesis is that those who consider *Architeuthis* to be a feeble swimmer perhaps tend to confuse rapid swimming with flexibility. The mantle locking apparatus is indeed poorly developed . . . and one can easily envision that, if an architeuthid were to "turn on a dime," it could well turn itself inside out. This is not to be confused with rapid swimming, but rather is indicative of a lack both of maneuverability and facility in changing direction with rapidity.

Moreover, wrote Aldrich, since *Architeuthis* is known to be one of the chief dietary items of sperm whales, it has to be able to escape: "The classical report on the speed of architeuthid swimming is that of Grønningsaeter (1946) . . . he clocked an architeuthid's speed at between 20–25 kn. If this observation is valid, and I believe it is, then the morphological apparatus with which the squid has been provided is clearly capable of speed sufficient to evade whales." (Sperm whales are capable of speeds of ten to twelve knots.)

When they examined three South African specimens of *Architeuthis,* Roeleveld and Lipinski found that the statocysts were proportionally large—no surprise there—but also that they were oriented obliquely, which suggests "that the natural position of *Architeuthis* may be at an oblique angle to the horizontal plane, with the head and arms, generally the heaviest part of the body, hanging downward." If this is the case, say the authors (following Pérez-Gándaras and Guerra), then the giant squid may hunt by using "some system of ambush, using the large tentacles to capture the prey." (Since every teuthologist has an opinion about the swimming and hunting capabilities of *Architeuthis,* Roeleveld and Lipinski weigh in with "the suggestion that *Architeuthis* is a poor swimmer and a passive and sluggish predator.")

In their 1996 study of the behavior of cephalopods, Hanlon and Messenger allocate a section to "giant squids" (including *Dosidicus, Moroteuthis, Taningia,* and *Mesonychoteuthis*), but this discussion is mostly devoted to the true giant of giant squids, *Architeuthis.* "Despite the size," they write, "the musculature of the mantle and arms is not particularly well developed. . . . The fins are not especially strong or large, and the 'giant' nerve fibers are much smaller than those of the common squid, *Loligo*. . . . Thus a picture begins to form of a relatively slow-moving, gentle creature, rather than an aggressive, shark-like monster."

In the spring of 1995, three architeuthids were trawled up off the west coast of Ireland. They were caught at a depth of approximately three hundred meters (984 feet) during a daytime tow near the bottom, suggesting that giant squid, like many other teuthids, rise toward the surface at night. All three Irish specimens were males, with a mantle length of about 3 feet. The first one, caught by the Dingle trawler *Shannon,* skippered by Michael Flannery, measured 20 feet to the tips of its tentacles, although it only weighed 60 pounds. All three had food in their stomachs, consisting of whiting, scad, prawns, octopuses, and other squids. In a 1997 letter to me, Colm Lordan of University College Cork wrote, "There is now a debate about whether giant squid scavenge food from the bottom or are active hunters. It now seems likely that they are active hunters, probably using their long tentacles to ambush their prey."

Roger Hanlon has summed up the behavior of *Architeuthis* in a way that seems to make a lot of sense.* First of all, its neutral buoyancy suggests that it is not an active hunter, but rather a "hoverer," holding its position at depth and waiting for its prey to come into the range of its long, grasping tentacles. Because those tentacles do not have the hooks of many of the other predatory squids, it probably doesn't grab large, powerful prey and subdue it. The small size of the suckers leads to the same conclusion. But since a half-ton squid probably needs about two hundred pounds of food a day, it also cannot be an eater of tiny plank-tonic animals, or even small fishes that it happens upon by accident. The Norwegian Torleif Holthe wrote in 1976, "Since it doesn't have any structures for catching small organisms, one must suppose that it takes prey that are so large that it can grab them with its tentacles and arms." (For an animal that lurks in the dark, one would expect to find some light organs—perhaps on the tips of the tentacles—that might attract the prey within range, but so far no one has found any evidence of bioluminescence in *Architeuthis*.) Its huge eyes, the largest eyes

---

*During the filming of a television show in June 1996 at Woods Hole, Hanlon and I had many opportunities to exchange ideas about the nature and behavior of *Architeuthis* and other teuthids.

ever known, may enable it to pick out its prey in conditions that would render most other creatures almost blind.

No one has seen a giant squid feeding—in fact, no one has ever seen a healthy giant squid doing anything at all—so until someone sees *Architeuthis* chasing something, the debate about its speed and strength will remain unresolved. Frederick Aldrich, probably the giant squid's most devoted advocate, assumed that *Architeuthis* is a powerful, aggressive animal, which at least in one instance was able to plow a great furrow in beach gravel before it expired. Precisely how the giant squid hunts—or what it eats, for that matter—is not known, but Aldrich believed that its prey consists mostly of sharks, rays, and skates, not the diet of a weak hunter.

Playing into this debate is the question of the giant squid's depth and range. Giant squid have appeared relatively consistently in the North Atlantic and antipodean waters, but we are now beginning to uncover much more about their habitat. The "unidentified animal" seen swimming offshore by road workers near Stinson Beach in 1983 might have been a giant squid (see page 29), but with nothing but hearsay reports, it is difficult to make an accurate identification. It would help, of course, if there were records of *Architeuthis* on (or off) the coast of California, and while none have been spotted on the beach, giant squid have certainly been documented in California waters. In 1971, Pinkas and colleagues identified a squid beak belonging to *Architeuthis* that was found in the stomach of an albacore caught off California. Further evidence that *Architeuthis* inhabits the eastern North Pacific is manifest in the appearance of its mandibles in the stomachs of sperm whales caught in California waters (Fiscus and Rice 1974), but recently more specimens have begun to appear, suggesting that this region might be—relatively speaking—densely populated with giant squid.

In 1980, 150 miles off the coast of southern California, an oceanographic sampling trawl was brought up with a 12-foot section of a giant squid tentacle attached. According to Robison (1989), "the tentacle had been wrenched from a living squid; the tissue was still elastic, the suckers contracted and gripped when they were touched, and the chromatophores contracted when rubbed." Since the squid could only have become entangled when the trawl was open, and since we know that depth, we know that the squid lost its tentacle at between five hundred and six hundred meters (1,640 to 1,970 feet), and we therefore have a faint clue about the depth range of the giant squid. As the author says, however, "the present data set is based on only a portion of a single *Architeuthis,* and it should not be extrapolated to represent whole body composition or species-wide depth habits. Unfortunately, much of what we know about this genus is based on such bits and pieces."

Perhaps the shroud of mystery is lifting. In a study published in 1985, Russian teuthologists Nesis, Amelekhina, Boltachev, and Shevtsov described a whole new crop of *Architeuthis* specimens that had been collected by "large-scale variable-depth trawlers, capable of capturing giant squid and verifying evidence as to the location and depth of the habitat." The first of these specimens was taken in September 1976 by the Spanish trawler *Yeyo* in South African waters at a depth of 375 meters—1,230 feet. (It was described in a 1978 paper by Spanish teuthologists Pérez-Gándaras and Guerra.) In 1980 and 1981, G. A. Shevtsov collected squids of various species from the Soviet research vessel *Novoulianovsk* (see note, pages 148–149) in the North Pacific. The largest specimen was collected in the open ocean, some fifteen hundred miles off Oregon, at a depth of only fifteen meters (about fifty feet) in water that was almost three miles deep. The mantle was 5.3 feet in length. The *Novoulianovsk* collected an astonishing *eighteen* more specimens in late March and early April 1980, some 250 miles off Los Angeles. Although they were considerably smaller than the monsters found in Newfoundland, Norway, or New Zealand—the mantles averaged only 2 feet in length—they were genuine giant squid of the heretofore gigantic (and formidable) genus *Architeuthis.*

In the southeastern Atlantic, A. P. Boltachev captured two more specimens. Working in waters some thirteen hundred feet deep off Zaire, the research vessel *Novoukrania* hauled in a giant squid. Unfortunately, the body was separated from the tentacles during the unloading process and fell overboard, leaving only the tentacles, which measured some 13 feet in length.

A couple of months later, *Novoukrania,* now farther south off Namibia, presented Boltachev with another *Architeuthis,* this one complete and measuring 14.46 feet in total length. Nesis, Amelekhina, Boltachev, and Shevtsov analyzed the existing data, and for the first time, they have been able to ascribe a distribution pattern to *Architeuthis.*

When a blue shark *(Prionace glauca)* was caught in the eastern equatorial Atlantic off Africa in 1972, portions of a large squid, identified as *Architeuthis* by the Russian teuthologist C. M. Nigmatullin, were found in its stomach. With a mantle length of 75 cm (29 inches), this specimen was hardly in the 50-foot class, but it was a large squid indeed. Using the figures employed by Nesis *et al.* (1985), where total length of some specimens was shown to be some 440 percent of their mantle length, this specimen's total length can be estimated at 10.6 feet. More interesting than its appearance in the stomach of a shark, however, is the location in which the shark—and therefore the squid—was captured. From the limited records, it was believed that *Architeuthis* bred in subtropical waters, and headed for the higher, colder latitudes to feed. With the exception of the juvenile specimen taken from the stomach of a lancetfish off Chile, no architeuthid has been found in tropical waters. This, however, might have more to do with limited sampling than absence of squid. Giant squid are known primarily from specimens that have been cast ashore or found in the stomach contents of sperm whales, and although nineteenth-century whalers often hunted their quarry "on the line," more recent sperm whaling has occurred in the colder waters of the North Pacific and the Antarctic. Nevertheless, this does not preclude juvenile squid inhabiting equatorial waters; as Nigmatullin wrote in his discussion of this specimen, "Possibly, in the future, when the mid-depth tropical fauna is better known, *Architeuthis* individuals will be found to be typical representatives."[*]

In September 1989, the Brazilian long-liner *Imaipesca* came across a carcass floating at the surface in water thirty-four hundred meters (11,200 feet) deep off the state of Santa Catarina. Brought to the Instituto de Pesca in São Paulo, the *lula gigante* was a female, with a mantle length of 5 feet and a weight of 200 pounds. According to Captain Hisami Funatsu, another piece of a large squid was seen in the stomach of a swordfish, and "another tuna boat operating in the same area observed a squid of large dimensions floating around the boat" (Arfelli *et al.* 1991).

The Russian scientists considered *Architeuthis* a "subtropical animal" that spawns and spends the first stages of its life in warmer waters, then migrates in

---

[*]Nigmatullin also mentioned "the discovery of large squids in the region under consideration," where Soviet fishermen in the eastern Atlantic observed large squids caught in tuna nets, or grasping the captured fish with their arms or tentacles. According to (unverified) verbal descriptions, they appeared to have been *Architeuthis.*

adolescence to feed in richer, colder waters near the polar fronts, or in regions where there are significant upwellings. With the advent of accurately calibrated midwater trawling devices, the vertical distribution of *Architeuthis* can now be postulated, and, of course, the more specimens that are examined, the more we are able to learn about the lifestyle and distribution of this heretofore enigmatic creature. "It is conjectured," wrote Nesis and his colleagues, "that their youth is spent at depths of 100–300 meters, whereas the adults live in mid-water and at the bottom, from 100–200 to approximately 500 meters. . . . At least during the warm part of the day, they seem to inhabit the epi- and mesopelagic zones, from subsurface levels to depths of at least 500–600 meters and probably deeper." The larger specimens—those with a mantle length of 10 to 16 feet—have been found only at the extreme boundaries of their known range, in the cold waters of Newfoundland, New Zealand, and Norway, but "the absolute majority of known types of *Architeuthis* had a mantle length of not more than 2–2.5 meters," suggesting that the mature individuals inhabit colder waters. (In those cases where the sex could be determined, the larger specimens were usually females.)

Even with this abundance of new material, the Russians could not resolve the problem of how many species of giant squid there are. They wrote, "More than a century of research concerning giant squid has thus not established with any certainty the number of species, not to mention any differences between species." There might be one species in each of the three isolated regions: *Architeuthis dux* in the North Atlantic, *A. martensii* in the North Pacific, and *A. sanctipauli* in the Southern Hemisphere. Or they might all be considered subspecies of *A. dux,* and therefore *A. dux martensii* and *A. dux sanctipauli.* But since so many of the specimens were small and possibly immature, we cannot discuss the taxonomy of the group with any degree of certainty. It is also possible that there is a pygmy or dwarf version, exemplified by the miniature, mature male with a mantle length of 6.5 inches, recovered from the stomach of a swordfish in Florida waters in 1978 (Toll and Hess 1981).

Despite the growing body of knowledge, *Architeuthis* is still a creature of mystery. It is difficult enough to envision a giant squid 55 feet long, and, given the traditional reluctance of these animals to show themselves, who can say that no larger ones remain undiscovered? All we can say with certainty about their maximum size is that none larger than 55 feet long has ever washed ashore. Moreover, the presence of *smaller* mature individuals casts the whole issue of *Architeuthis*-as-monster into question. The giant squid continues to be a giant enigma. In her keynote address to the American Institute of Biological Sciences in February 1991, Sylvia Earle, one of America's premier oceanographers (and the women's world record holder for deep diving), said:

> And what of that never-never creature, that squid of all squids, *Architeuthis dux,* the giant squid? There are many sea stories, numerous possible glimpses, and fragments of animals taken from the stomachs of sperm whales or washed up on various beaches. But no one has recorded certain, direct observations of these creatures in their own realm. It would not be likely that an animal eighteen meters long would escape notice in any terrestrial habitat on land, but it has been possible thus far for giant squids to elude even highly motivated ocean scientists.

Humans may not know very much about the habits of *Architeuthis,* but there is a group of large-brained, warm-blooded mammals that knows a great deal. (I apologize for the anthropomorphic use of "know" in this context; we have no idea what—or even if—animals "know.") The sperm whale, known to scientists as *Physeter macrocephalus,* is known to the giant squid as its archenemy. In fact, sperm whales might be described as the archenemies of all squids, just as we would describe lions as the enemies of zebras, or great white sharks as the enemies of sea lions: they eat them when they can catch them.

Battle
of the
Giants

While scientists were debating the existence of the kraken, another group of men knew that there was certainly something that fit the early descriptions, and, in fact, many of them had actually seen it. Yankee sperm whalers, plying the world's oceans in pursuit of the cachalot, often saw their prey, in its death throes, regurgitate large pieces of *something,* and they often hooked one of the pieces to get a closer look at it. In 1856, Charles Nordhoff wrote a book called *Whaling and Fishing,* in which he described what the whalemen called "squid," but which he believed to be "a monster species of cuttle-fish":

> The animal seldom exhibits itself to man; but pieces of the feelers are often seen afloat, on good whaling ground. I have examined such from the boats, and found them to consist of a dirty yellow surface, beneath which appeared a slimy, jelly-like flesh. Of several pieces which we fell in with at various times when in the boats, most had on them portions of the "sucker," or air exchanger with which the common cuttle-fish is furnished, to enable him to hold the prey about which he has slung his snake-like arms. These floating pieces are supposed to have been bitten off or torn by the whales, while feeding on the bottom. Many of those we saw were the circumference of a flour barrel. If this be the size of the arms, of which they probably have hundreds, each furnished with air exhausters the size of a dinner plate, what must be the magnitude of a body which supports such an array?

Of all the well-known predator-prey combinations, none are more intimately connected than the sperm whale and the giant squid. The sperm whale preys on other teuthids, but it is likely that nothing preys on full-grown giant squid except the sperm whale. Giant squid have often been accused of causing the circular scars on the heads of some sperm whales, either in an attempt to eat the whale or, more likely, in the struggle not to be eaten. (Interviewed by Harry Thurston for a 1989 article in *Equinox,* Frederick Aldrich was quoted to the effect that in a battle with a sperm whale, the giant squid always loses.) Though many believe that

Architeuthis is a sluggish animal, neither powerful nor aggressive, such suppositions are never going to interfere with the creature's enduring reputation as a man-eating, ship-grabbing, whale-wrestling monster. It has long been a part of sperm whale lore that the whale has to dive to prodigious depths to seek out the giant squid, its favorite prey. The size and alleged ferocity of *Architeuthis* provide more than enough ammunition for the creative writer or overimaginative naturalist, and titanic battles between the squid and its archpredator, the sperm whale, will continue to appear.

In *Moby-Dick,* as the *Pequod* sails northeast of Java, the harpooner Daggoo saw a "strange spectre" from his perch on the main masthead:

> In the distance, a great white mass lazily rose, and rising higher and higher, and disentangling itself from the azure, at last gleamed before our prow like a snow-slide, new slid from the hills. Thus glistening for a moment, as slowly it subsided, and sank. Then once more arose, and silently gleamed.

They lowered the boats and "gazed at the most wondrous phenomenon which the secret seas have hitherto revealed to mankind":

> A vast pulpy mass, furlongs in length and breadth, of a glancing cream color, lay floating on the water, innumerable long arms radiating from its centre, and curling and twisting like a nest of anacondas, as if blindly to clutch at any hapless object within reach. No perceptible face or front did it have; no conceivable token of either sensation or instinct; but undulated there on the billows, an unearthly, formless, chance-like apparition of life.

Herman Melville's massive, mysterious novel is generally considered the consummate achievement of American literature, the Great American Novel. It has been called an elegy to democracy, a tract on the nature of religion, an investigation of man's relationship to the natural world, a conflict between the eternal forces of good and evil. Because Melville knew the whale fishery firsthand (he shipped out on the New Bedford whaler *Acushnet* in 1837), *Moby-Dick* also includes the best description ever written of nineteenth-century Yankee whaling. It is fiction, of course, but the accoutrements of whaling and natural history that Melville wove into his story are as accurate as any nineteenth-century biologist's—and a lot better written. Thus, if the whalemen knew anything about

At the conclusion of the chapter called "Squid" in the 1930 edition of Moby-Dick, *Rockwell Kent drew a ghostly apparition in the sky.*

giant squid, Melville recorded it in *Moby-Dick.* He wrote: "By some naturalists who have vaguely heard rumors of the mysterious creature, here spoken of, it is included among the class of cuttle-fish, to which, indeed in certain external respects it would seem to belong, but only as the Anak of the tribe."*

The Reverend Henry Cheever, who sailed aboard the whaler *Commodore Preble,* wrote a book in 1859 titled *The Whale and His Captors.* Most of the "captors," of course, were men with harpoons, but he did present an account (not seen by him, but "sworn before a Justice of the Peace in Kinebeck, Maine, in 1818") of what appears to have been an encounter between a sperm whale and a giant squid:

> The serpent threw up its tail from twenty-five to thirty feet in a perpendicular direction, striking the whale by it with tremendous blows, rapidly repeated, which were distinctly heard, and very loud, for two or three minutes; they then both disappeared, moving in a south-west direction; but after a few minutes reappeared in-shore of the packet, and about under the sun, the reflection of which was so strong as to prevent their seeing so distinctly as at first, when the serpent's fearful blows with his tail were repeated and clearly heard as before. They again went down for a short time, and then came up to the surface under the packet's larboard quarter, the whale appearing first, and the serpent in pursuit, who was again seen to shoot up his tail as before, which he held out of water for some time, waving it in the air before striking, and at the same time his head fifteen or twenty feet, as if taking a view of the surface of the sea.

---

*In the Old Testament (Numbers 13:33), Anak is the father of a race of giants that threaten the Israelites during their exodus from Egypt.

After being seen in this position a few minutes, the serpent and whale again disappeared, and neither was seen after by any on board. It was Captain West's opinion that the whale was trying to escape, as he spouted but once at a time on coming to the surface, and the last time he appeared he went down before the serpent came up.

Since one cannot actually see the head of *Architeuthis* when the animal is in the water, it seems obvious that both the "head" and the "tail" of this creature were the tentacles of a giant squid, assigned to either end of a sea serpent.

If the battle comes to the surface and there are people to observe it, we might get some idea of what the struggle really looks like. On July 8, 1875, the crew and officers of the bark *Pauline* saw a battle between a sperm whale and *something*. As later reported in the *Illustrated London News*,

In the 1930 edition of Moby-Dick, the artist Rockwell Kent drew this strange illustration of a giant squid at the surface.

Captain George Drevar was some twenty miles off Cape San Roque (Brazil) when the *Pauline* came upon "a monstrous sea serpent coiled twice round a large sperm whale." The "sea serpent" conquered the whale in Drevar's narration, pulling the hapless cetacean below the surface, "where no doubt it was gorged at the serpent's leisure." A week later, Captain Drevar was still in the same latitude, now eighty miles from shore, when he "was astonished to see the same or a similar monster. It was throwing its head and about 40 feet of its body in a horizontal position as it passed onwards by the stern of our vessel." Drevar described the serpent's mouth as "always being open," but with this exception, the creature sounds suspiciously like a giant squid. (Heuvelmans, ever on the lookout for verifiable sea serpents, concludes that the open mouth and the coloration cannot be reconciled with a squid, and writes that it was a snake, or more likely, a giant eel, which, he claims, "are immensely powerful constrictors"; but they certainly are not. [1965])

Frank Bullen (1857–1915), like Melville a whaleman-turned-author, had no compunctions about describing a clash between a squid and a whale. Bullen's narrative, *The Cruise of the "Cachalot,"* is replete with improbably theatrical episodes of courage—usually his own—and wildly unlikely behavior on the part of various animals, and most historians are inclined to dismiss his book as more fiction than fact. Nevertheless, the book does contain a vivid description of a battle between the two gigantic predators, a portion of which is here reproduced:

> A very large sperm whale was locked in deadly conflict with a cuttle-fish, or squid, almost as large as himself, whose interminable tentacles seemed to enlace the whole of his great body. The head of the whale especially seemed a perfect network of writhing arms—naturally, I suppose, for it appeared as if the whale had the tail part of the mollusc in his jaws, and, in a business-like, methodical way, was sawing through it. By the side of the black columnar head of the whale appeared the head of the great squid, as awful an object as one could well imag-

In 1875, viewers aboard the bark Pauline *off Brazil saw what some described as a fight between a sea serpent and a sperm whale. Except for the "eye," it is quite easy to see the "serpent" as the tentacle of a giant squid.*

ine even in a fevered dream. Judging as carefully as possible, I estimated it to be at least as large as one of our pipes, which contained three hundred and fifty gallons; but it may have been, and probably was, a good deal larger. The eyes were very remarkable from their size and blackness, which, contrasted with the livid whiteness of the head, made their appearance all the more striking. They were, at least, a foot in diameter, and seen under such conditions, looked decidedly eerie and hobgoblin-like.

Bullen's narrative (which was published in 1898) appears to owe something to Melville's, especially as concerns the coloration of the squid. (In Melville's novel, the living squid is a ghostly white color and the harpooner Daggoo actually mistakes it for Moby Dick, crying out, "There! there again! there she breaches! right ahead! The White Whale, the White Whale!") The amount of salt required to season Bullen's story is evident from his setting: the cuttlefish and the whale staged their epic battle by moonlight.

When Victor Scheffer, a respected marine biologist, wrote of an encounter between a sperm whale and a giant squid in *The Year of the Whale*, he had the full weight of science behind him. Nevertheless, he described an encounter that neither he nor anyone else has ever witnessed:

The pressure is now one hundred tons to the square foot; the water is deathly cold and quiet. At a depth of three thousand feet he levels off and begins to search for prey. The sonar device in his great dome is operating at full peak. Within a quarter-hour he reads an attractive series of echoes and he quickly turns to the left, then to the right. Suddenly he smashes into a vague, rubbery, pulsating wall. The acoustic signal indicates the center of the Thing. He swings open his gatelike lower jaw with its sixty teeth, seizes the prey, clamps it securely in his mouth, and shoots for the surface. He has found a half-grown giant squid, thirty feet long, three hundred pounds in weight. The squid writhes in torment and tries to tear at its captor, but its sucking tentacles slide from the smooth, rushing body. When its parrot beak touches the head of the whale it snaps shut and cuts a small clean chunk of black skin and white fibrous tissue. . . . He crushes the squid's central spark of life, its gray tentacles twist and roll obscenely like dismembered snakes. . . . Now at ease, the bull turns the dead beast and leisurely chomps it into bite-sized pieces, each the size of a football, and thrusts them mechanically into his gullet with his muscular tongue.

Except for the part where the whale chomps the squid into bite-sized pieces (Victor Scheffer would know that the whale's jaws are unsuited for chomping things into pieces), this is probably what it looks like when a whale catches a squid. The squid might struggle a little more, and it has more than "sucking tentacles" to grip the skin of the whale: it has sharp-toothed suckers that can dig into the whale's sensitive skin.

Giant squid and sperm whales are known to inhabit Norwegian waters; a colossal conflict is a distinct possibility. In *Norges Dyreliv* (Animal Life of Norway), Einar Koefoed relates a tale about the battle of the titans:

> A Norwegian whaler once saw a large whale with the body of a giant squid between its teeth. . . . The giant squid had thrown its tentacles, as thick as ropes, around the head of the whale so that it could not open its jaws. Suddenly the squid let go and the whale submerged. After a while it surfaced again with the crushed squid in its mouth.

There are very few contemporary accounts of a battle between a giant squid and a sperm whale, but just before the outbreak of World War II, J. W. Wray was sailing off the Kermadec Islands (north of New Zealand) when he spotted a disturbance in the water. As he tells the story in *South Sea Vagabonds,* he sailed over to investigate "when, to our amazement, a giant tentacle, easily twenty-five feet long, came out of the water, waved around for a second, and crashed back into the water again. A few seconds later our hearts stood still as the fore part of a large whale shot out of the water barely thirty yards away and, encircling its head, its enormous tentacles thrashing the water, was a truly villainous giant octopus. . . . Two more giant tentacles emerged, making a terrific commotion on the water." Although Wray calls the animal an octopus, the length of the tentacles makes the case for a giant squid, and the battle with the whale (unidentified, but occurring in an area well known for sperm whales) certainly suggests the traditional antagonists.

In 1887, the Danish zoologist August Fjelstrup published a paper in which he described various round markings on the heads of pilot whales as the rudiments of vibrissae, or whiskers. Fjelstrup also quoted Frederick Debell Bennett's 1840 *Narrative of a Whaling Voyage Round the Globe,* in which Bennett wrote of the blackfish (the pilot whale), "On the head and chiefly around the lips the skin is marked with many scattered circles, each the size of a sixpence and composed of a single row of small, depressed dots which would appear to mark a disposition to the formation of vibrissae." Apparently many nineteenth-century zoologists agreed with Fjelstrup and Bennett, and even though no pilot whale had ever been found with vibrissae, there was no better explanation.

Because they are mammals, many whales have hairs, usually on the face, at some stage of their lives. Fin whales have some hairs at the apex of the lower jaw, and those "bumps" on the face of humpbacks, which have been likened to stove

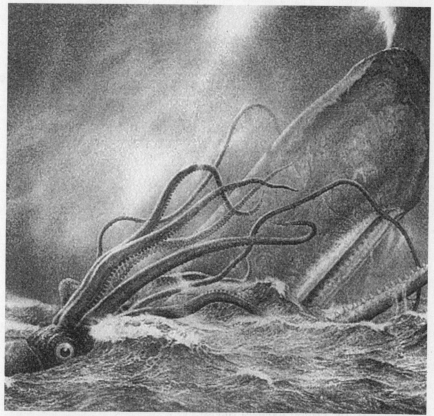

Encounters between these two giants are more likely to take place in the depths, but in order to compare the adversaries, Glen Loates has drawn the sperm whale and the giant squid at the surface.

bolts, are actually hair follicles. But neither pilot whales nor sperm whales, the animals that have those mysterious circular scars around their mouths, have hairs growing out of the circles. Another Dane came up with a more realistic answer than whales with whiskers. Adolf Jensen of the University of Copenhagen wrote a little piece called "On Some Misinterpreted Markings on the Skin of the Caaing Whale,"* in which he observed that since the pilot whale is a known consumer of squid, the marks obviously "represent impressions in the whales' skin caused by the suckers of ten-armed cuttlefish." Jensen identified the makers of the marks as *Architeuthis, Onychoteuthis,* and also *Ommastrephes,* even though he recognized that pilot whales normally feed on the smaller, rather than the larger, species of squid.

Captured sperm whales are often scarred with round marks that look very much like the marks that would have been left by the suckers on the arms of giant squid. Since it is known that these whales eat squids of all sizes (but mostly smaller than *Architeuthis*), it is not an unreasonable assumption that these scars are caused by squid unwilling to be eaten by whales.

In "Body Scarring on Cetaceans—Odontocetes," a paper published in the *Scientific Reports of the Whales Research Institute* in 1974, Charles McCann proposed several theories about the scars on the heads of various whales. He also included some rather bizarre observations about sperm whales and squids, which, as far as I know, have not appeared elsewhere. He wrote:

> The enormous head of the sperm whale occupies almost one-third of the
> total length of the animal. It is provided with the cushion of spermaceti, a waxy
> looking substance already referred to. To speculate, this "cushion" functions as
> a shock-absorber when the animal uses its head as a battering ram in offence and
> defence. In addition, the array of large teeth serve as weapons of defence and
> offence, and they function as grasping organs in dealing with the large slithery

---

*In various northern countries, the pilot whale (*Globicephala* spp.) was known as the *ca'ing, caa'ing,* or *caaing* whale. The origin of this strange name is unclear, but it seems to have something to do with "calling." Sir Francis Fraser, esteemed cetologist of the British Natural History Museum, disagreed, and wrote, "It may be stated that the Scottish name caa'ing whale is connected with the practice of driving or herding the animals and has nothing to do with calling or vocalization of any sort" (Norman and Fraser 1938).

cephalopods (particularly the Giant Squid) which form a large proportion of the food of Sperm Whales.

In dealing with some of the larger cephalopods, there seems to be a suspicion that the Sperm uses its head to pummel the squids after grasping them with its jaws (the fleeing squid, going backwards as is its wont, would be grasped by the tentacles). Evidence of battles with large cephalopods is impressed on the facial region of the Sperms.

The image of a sperm whale grasping a squid by the tentacles and bashing it on the bottom is indeed an unusual one, as is the picture of the whale "pummeling" the helpless squid with its battering-ram head. We still do not know precisely how the whales catch the squids, but McCann's "speculations" are too absurd to be taken seriously.

In *Depths of the Ocean* (1912), Johann Hjort made one of the most often quoted—and misquoted—remarks about the giant squid. In 1902 Hjort was aboard the research vessel *Michael Sars* in the North Atlantic when they came across a small giant squid floating dead at the surface north of the Faeroes. There was nothing spectacular about this sighting, but then Hjort wrote, "In 1903 in Iceland I had the opportunity of making an interesting observation, showing the gigantic dimensions of these squids." The ship visited the whaling station at Mjofjördur, where there were two freshly killed whales, a sperm whale and a right whale. Hjort wrote:

> Inspecting the cachalot, I saw around its enormous jaw several long parallel stripes, consisting, as closer scrutiny revealed, of great numbers of circular scars or wounds about 27mm in diameter. . . . It occurred to me that these scars must have been left by the suckers of giant squid, and following up this idea I found in the whale's mouth a piece of squid tentacle 17cm in maximum diameter.

In Hjort's book there is a picture of the "skin of the cachalot with marks from the struggle with *Architeuthis*. Nat. size." The scars that Hjort measured at 27 mm were a little over an inch in diameter, and in the photograph the largest one measures 1 inch across. The picture has been reproduced in almost every discussion of sperm whales and squid, but somehow, the diameter of the circular scars has increased to impossible proportions, perhaps through a confusion of the 27 mm scars and the 17 cm diameter *tentacle,* which equals 6.63 inches.*

According to Clarke (1980), the largest circular scars on the whales' heads have come from *Architeuthis,* the squid with the largest suckers. He wrote, "I have not yet seen conclusive evidence to suggest that *sucker* scars are larger than 3.7 cm [1.44 inches] across." In 1872, the Reverend Harvey measured the suckers of the Bonavista Bay specimen at 2.5 inches in diameter, and in his 1938 monograph on sperm whales, L. Harrison Matthews wrote, "Nearly all male Sperm whales carry scars caused by the suckers and claws of large squids, scars caused by suckers up to 10 cm. in diameter being common. The claw marks take the form of scratches 2–3 m. in length, and appear to be of more frequent occurrence than sucker marks."†

Despite the dramatic image of gigantic cephalopods with vicious talons on their tentacles, the "claws" of the large squids are not nearly as frightening as the writers of teuthid fiction would have us believe. Many of these claws are sheathed in modified suckers, since they "always develop from suckers by the uneven elongation, bending, and longitudinal folding of one or (rarely) two teeth [on the rim of the sucker]" (Nesis 1982). In the original description of *Mesonychoteuthis* (from the two specimens found in the stomachs of South Shetlands sperm whales), G. C. Robson described the largest hook as 2.5 cm (just under an inch) in length, by about a quarter of an inch wide.‡ In his description of a large specimen of *Moroteuthis* that washed ashore at Unalaska in the Aleutians in 1896, D'Arcy

---

*A few of the more egregious examples follow: in the Time-Life book *Dangerous Sea Creatures* (written by T. A. Dozier) we read that "an ordinary giant squid of 50 feet leaves teeth-ringed sucker marks measuring between three and four inches across on a whale, but sperm whales have been captured with tentacle marks 18 inches across." In *The Guinness Book of Animal Facts and Feats,* Gerald Wood wrote that scars "measuring up to 5 in. in diameter have been found on the skins of sperm whales captured in the North Atlantic," and Willy Ley—who should have known better—wrote (in *Exotic Zoology*), "Another claim goes for marks on the skin of such a whale of a sucking disk over 2 feet in diameter."

†Matthews's measurements of 10 cm—3.9 inches—are so much larger than any other recorded sucker dimensions that one suspects some sort of error, either in measuring or in transcription.

‡When I wrote to Kir Nesis in Moscow to ask him about the claw sizes of various squid species, he answered me by saying, "But 20–25 mm is not so small. Dr. Yulia Filippova, a known Russian squid specialist of the Russian Federal Research Institute of Fisheries and Oceanography (VNIRO) has a necklace of *Mesonychoteuthis* hooks—very impressive!"

Thompson wrote of the hooks, "the largest, toward the middle of the club, are about five-eighths of an inch long, and with bases five-sixteenths of an inch broad."

Architeuthis has no claws, but other large squids do. Indeed, some of the most popular items in the whales' diet are equipped with talons, and the reluctance to be swallowed of such species as *Moroteuthis, Mesonychoteuthis,* and *Taningia* may provide an explanation for some of the scratches.

At a known maximum length of 19 feet, *Moroteuthis robustus* is one of the largest of all squids. (There are at least four other species in the genus *Moroteuthis: M. ingens, M. robsoni, M. loennbergii,* and *M. knipovitchi,* none of which gets quite as big, and all of which are characterized by the presence of hooks or claws on the tentacular clubs.) There is also an Antarctic squid known as *Kondakovia longimana,* which is similar to *Moroteuthis,* but where *Moroteuthis* has only hooks on the tentacle clubs, *Kondakovia* has hooks and suckers.

Photograph of the skin of a sperm whale marked with scars made by the suckers of squid. In the original, the circles are about 1 inch in diameter, but the size is usually exaggerated to suggest that squid suckers get much larger.

Moroteuthis robustus has no common name, although it is sometimes called the Pacific giant squid. (Roper, Sweeney, and Nauen's *Cephalopods of the World,* Volume 3 of the *FAO Species Catalogue,* gives the awkward—and probably never used—name "robust club-hook squid.") In addition to the tail fins, *M. robustus* also has a pointed "tail" that extends beyond the fins. Its range covers the North Pacific arc from California to Japan, and includes Alaska and the Aleutians.* *Architeuthis* is also found in these waters—its range map in the *FAO*

---

*In Japan, *Moroteuthis* is known as *nyudoo-ika,* which can be roughly translated as "monk [or

Catalogue has all the temperate and tropical waters of the world shaded in—but records for the North Pacific are sparse. Therefore, a very large squid recorded from this area is much more likely to be *Moroteuthis* than *Architeuthis*. When W. H. Dall of the Smithsonian described some Aleutian cephalopods in 1873, he wrote that a number of "giant cuttles" had washed ashore at Unalaska. One, he wrote, was "perhaps *Onychoteuthis bergi,* a specimen of which measured from the posterior end of the body to the mutilated ends of the tenticular arms one hundred and ten inches with a body girth of nearly three feet and weighing nearly two hundred pounds." There is a squid known as *Onychoteuthis boreal-ijaponica* found in this area, but it reaches a maximum mantle length of 15 inches, so it is more likely that the species described by Dall was *Moroteuthis*.

Within its North Pacific range, *Moroteuthis robustus* is known as a favorite food of sperm whales, because the same area of the North Pacific was the favorite hunting ground of Japanese and Russian sperm whalers in the mid–twentieth century, and they had more than ample opportunity to examine the stomach contents of the whales they brought aboard their factory ships.* In a study of the squids retrieved from the stomachs of sperm whales of the Bering Sea and the Gulf of Alaska, Okutani and Nemoto found that *Moroteuthis* was the favorite food of sperm whales in those waters. They wrote, "Because of its huge size, reddish variegation and rippled skin, this species is easily discriminated at the field observation on board the factory ship. One of the specimens brought home was measured at about 700 mm [27 inches] in total dorsal length." (*Huge size* seems to be a relative term.)

The Pacific giant squid is believed to live near the bottom at depths that range from two hundred to six hundred meters (650 to 2,000 feet), although Hochberg and Fields report that it is "occasionally seen swimming at the surface or stranded at the surf line."

---

man-with-a-shaved-head] squid," perhaps from some perceived resemblance of the mantle to the tonsure of a monk. (See pages 61–65 for a discussion of the *Monachus marinus,* or "Sea Monk" from European waters.)

*The heyday of the commercial whaling industry was not, as many people believe, the days of square-rigged whalers out of Nantucket and New Bedford, but in the 1960s, when the Soviets and the Japanese killed more sperm whales *every year* than the entire Yankee whaling fleet took in its century-long massacre. In 1965, twenty-three thousand sperm whales were killed in the North Pacific, while the Yankee fishery was not thought to have killed more than twenty thousand whales in its entire history (Townsend 1935).

"The stomachs of sperm whales taken off California and British Columbia," wrote Dale Rice in 1978, "contain mostly the large squid *Moroteuthis robustus*. I have measured squid specimens that were up to four feet five inches from the tip of the tail to anterior edge of the mantle and eleven feet four inches to the tip of the tentacles." Like *Architeuthis,* this species has substituted ammonium ions for the sodium ions in its muscle tissue, which accounts for its low density and its bitter taste. (Sperm whales don't seem to mind a little taste of ammonia; some of their favorite foods, such as *Architeuthis* and *Moroteuthis,* reek of it, but since sperm whales have neither taste buds nor a sense of smell, a little ammonia isn't going to affect their eating habits.) In fact, the ammonium content (but not the actual taste) of these squids may be one of the factors that contribute to their popularity in the sperm whales' diet. The ammoniacal squids are neutrally buoyant, and can therefore remain motionless in the water column. The stationary behavior of the squids may somehow be related to the hunting techniques of the whales. (It has to be easier to catch something that is standing still than something that is darting hither and yon.)

During the 1963–64 sperm-whaling season in New Zealand, 133 carcasses were examined by cetologist David Gaskin. He found that squids of the genus *Moroteuthis* (probably *M. ingens,* a large Southern Ocean form that reaches a mantle length of 3 feet) made up 74.8 percent by weight of the fresh squid species in the sperm whales' stomachs. As they crossed the flensing deck at the Tory Channel Whaling Station, Gaskin and Martin Cawthorn noticed that the squids "were glowing a bluish white, and that the light was visible for many yards." Since the luminescence "came away in the hands," and was therefore not produced by specific light organs, they realized that it was a "property of the mucous covering of the squid or . . . unicellular organisms held in suspension." Many deep-sea creatures—siphonophores, echinoderms, jellyfish, and even fishes—can produce a luminescent mucus, but "generalized luminescence of the kind recorded by the present author [Gaskin 1967] does not seem to have been recorded previously for squid in available literature."*

---

*According to Nesis (1982), "The skin of cephalopods is thin but is very complicated in nature. The upper layer—the epidermis—is formed by a single layer cylindrical epithelium with numerous mucous cells. The mucous makes the body of the cephalopod slimy, which makes movement easier in the water. Under the epidermis lies connective tissue containing muscular tissue, chromatophores, and iridocytes."

Gaskin's article was written some time before the development of the various theories about sperm whale feeding techniques, such as that the whale emits sonic blasts to stun the squids or, more recently, that the whale might use sound to light up the depths so it can see what it is hunting. He probably knew about Beale's idea that the squid might be attracted to the white lower jaw of the whale (see page 159) when he wrote,

> The author has always considered it unlikely that such a large animal as a full grown male sperm whale, weighing perhaps 50 tons, could conceivably chase separately every squid eaten. Considerable energy must be expended swimming down to depths of perhaps 500 fathoms, without taking into account actually chasing squid, which are fast-moving animals. The long narrow shape of the lower jaw of the sperm whale does not seem compatible with the capture of fast-moving prey at speeds of several knots. The jaw has little lateral movement.

Given the speed of the squids and the whale's difficulty in capturing them, Gaskin suggests that the idea of "active feeding" might have to be dropped, and a "plausible method of passive feeding suggested":

> The luminescent mucous described above could easily be transferred to the lining of the sperm whale's mouth and act as a lure to attract more squid. If the mucous collected on or between the mandibular teeth the regular spacings might at a distance give the appearance of a fish or other animal with light organs along its flanks. Predatory fish or squid could be attracted to the light, and the sperm whale could lie almost motionless in the water and swallow the animals as they came to its mouth. The whale might have to capture a few squid actively before enough mucous accumulated.

It is an ingenious construct, but one that might be a little difficult to prove. Besides, it presupposes that sperm whales throughout the world's oceans feed only on mucus-covered bioluminescent squids. What happens when the "passive" whale encounters a squid species that has no mucus?

In a 1959 article in *Sea Frontiers* called "Hunting Sea Monsters," Gilbert Voss mentioned "a squid that could qualify in the most lurid deep-sea drama. While the body was small and weak, the head bore enormous jaws and was surrounded by eight heavy muscular arms which bore great hooked claws and two long ten-tacles." He wrote that Dr. Anna Bidder of Cambridge "has other remains which point to a squid about 24 feet long, and which may grow even longer. It is no relation to *Architeuthis* but is fed upon by the sperm whales of the Southern Ocean." Dr. Bidder never got around to publishing the description, but others did. It was *Mesonychoteuthis hamiltoni,* one of the most formidable of all the sperm whale's menu items.

As described by Roper, Sweeney, and Nauen, *Mesonychoteuthis* is "a very large species" that figures prominently in the diet of Antarctic sperm whales. According to G. C. Robson's 1925 description of the type specimen (based on fragments of two specimens collected from the stomachs of sperm whales taken off the South Shetland Islands), the longest arm was 46.3 inches long, and its "hand" was equipped with a series of swivel-based hooks that could be rotated in any direc-tion. The name can be translated as "middle-hooked squid," and refers to the

Mesonychoteuthis hamil-toni *is an Antarctic species that can reach a total length of 20 feet. This specimen, photographed by Alexander Remeslo, was taken in 1981 at approximately twenty-five hundred feet in the Antarctic off Droning Maud Land by the Soviet trawler* Evrica.

location of the double row of hooks on the middle of each arm, between the basal and terminal ringed suckers. This is the only large squid with hooks on the tentacles as well as on the arms. The body of *Mesonychoteuthis* can be as large or larger than that of *Architeuthis*, but its tentacles are much shorter. Nesis refers to it as a "giant," with a mantle length of 200 to 225 cm (7 to 7.5 feet) and a total length of 350 cm (11.37 feet), not including the tentacles, and one was caught in the Antarctic in 1981 that was nearly 17 feet long. *Mesonychoteuthis* is believed to have a circumpolar Antarctic distribution, and according to Nesis (1982), it is "the leading member of the Antarctic teuthofauna by biomass."

Hardly anything is known of the biology of *Mesonychoteuthis;* like *Architeuthis,* it has never been seen alive, and almost all of our information comes from the examination of dead (and often semidigested) specimens. Its tentacles are comparatively short and thick, and its tail fins are broad, muscular, and heart-shaped, accounting for its name in Japanese, *dai-oo-hoozukai-ika.* As in *dai-oo-ika,* the name for *Architeuthis, dai-oo* is "great king," *ika* is "squid," and *hoozukai* is the bladder cherry plant, familiarly known as the "Chinese lantern," whose calyx is a fragile, bright orange bag, pointed at the end away from the stem.

Mesonychoteuthis is a cranchid squid, characterized by the fusion of the mantle to the head at one ventral and two dorsal points, and by photophores on the ventral surface of the prominent, sometimes protruding, eyes. Another large cranchid is *Galiteuthis,* a well-armed squid with large, sheathed hooks on the tentacles. In her 1980 review of the Cranchiidae, Nancy Voss describes this species as "moderately large . . . long, slender, broadest in anterior half, tapering posteriorly to slender point; mantle wall thin, muscular, and, in subadults or adults may or may not bear numerous cartilaginous tubercles over outer surface." Nesis (1982) wrote that there are "five species [of *Galiteuthis*] in bathypelagic and bathyal zones . . . of all oceans, except the Arctic Ocean. Mantle length to 66 cm [26 inches] but probably to 2.7 m [8.85 feet]."

The Russian vessel *Novoulianovsk,** working in the Sea of Okhotsk in 1984, brought up the remains of a gigantic specimen of *Galiteuthis phyllura* from a

---

Novoulianovsk is a large stern-trawler that was sent on oceanographic research expeditions from 1980 to 1992. According to Nesis (personal communication), she had mighty engines and a trawling winch capable of towing the largest commercial trawls at speeds up to five knots. She towed huge, midwater trawls of two thousand feet in length, at depths up to eighteen hundred

depth of one thousand to thirteen hundred meters (thirty-three hundred to forty-three hundred feet), and Nesis (1985) said that it was "almost as large as *Mesonycho-teuthis hamiltoni* (of the same family)." Only an arm and a tentacle were collected, but they were so large (the arm was 40 cm long [15.6 inches] and the tentacle 115 cm [44.8 inches]) that Nesis was able to estimate the mantle length at 265 to 275 cm (8.61 to 8.93 feet), and the total length at over 4 meters (more than 13 feet). "Because of its narrow body," wrote Nesis, "we conclude that its mass is consistently lower than that of the other large squids." In a popular publication Nesis called this species the "deep-sea leaf-tailed squid, since the leaf-like tail is a characteristic feature of the genus *Galiteuthis*."

On exhibit in the National Museum of Natural History (Smithsonian Institution) in Washington, there is a specimen of the squid known as *Taningia danae,* 7 feet long and weighing 135 pounds. Unlike most other squids, *Taningia* does not have two long feeding tentacles, but it does have something more surprising: on the ends of two of its arms are gigantic yellow photophores, the largest light-producing organs in any known animal. These lemon-colored (and lemon-sized) photophores can be flashed at will, because they are equipped with a black, eyelidlike membrane that can be opened or closed. In addition, this species has claws—likened to those of a cat—on the suckers of its arms. The Smithsonian's specimen was dumped into the hold of the Georges Bank fishing boat *Defender* as Captain George Dow emptied his nets. He was two hundred miles southeast of Portland, Maine, when his engineer came up to the bridge and asked, "Ever seen a seven-foot squid?"

Dow brought the squid aboard, iced it down, and delivered it to the National Marine Fisheries Service (NMFS) laboratory in Gloucester, Massachusetts, where it was measured and photographed. Surprisingly, this was the first record of *Taningia* from the western North Atlantic, one of the most heavily fished areas in

---

meters (fifty-nine hundred feet). "Most squids," he wrote, "lose tentacles (they are torn out) but there are many tentacles and arms without squids. Such was the case with my giant *Galiteuthis* with the body missing and only the 'hoofs and horns' remaining. . . . The bottom trawl is more fine-meshed throughout, and bottom squids and octopuses come to the deck in much better form, some alive. But in all cases, my job was mostly to crawl along the deck, pincers in hand, and pick out the squids and their details like a nuthatch running along the tree's trunk." Ulianovsk is a large city on the Volga River, and Novoulianovsk is its industrial suburb.

the world, and one in which biological sampling has been going on for well over a century.

Not much is known of the biology of *Taningia*, but sperm whales must have some of the answers, since most known specimens have been taken from the stomachs of captured whales. (According to a 1993 review by Roper and Vecchione, other specimens have come from the stomachs of sharks, lancetfishes, tunas, wandering albatrosses, and elephant seals.) In 1959, the stomach contents of a whale were examined by Malcolm Clarke at the Canaçal whaling station on the island of Madeira. There were four thousand beaks, twenty-eight partially digested squids, and "a perfectly intact specimen" of the squid then known as *Cucioteuthis unguiculata,* but now called *Taningia danae.* ("Other specimens," wrote Clarke, "[that] were referred to *Cucioteuthis unguiculata* should all be regarded as *Taningia danae.*") Clarke (1962b) wrote that it was "easily recognized by the large, gelatinous body, the broad fins extending almost to the front of the mantle, the absence of long tentacles (there being only eight arms); and the presence of strong hooks on the arms." The mantle was measured at 140 cm (4.55 feet).

From the records collected by Roper and Vecchione, it appears that *Taningia* has an almost worldwide distribution, having been collected—largely from the stomachs of sperm whales—in the western North Atlantic and off Bermuda, Hawaii, South Georgia, South Africa, Japan, Australia, New Zealand, the Azores—in short, almost everywhere that sperm whales were hunted.* They wrote, "With the addition of the material in this paper, the geographical distribution of *Taningia danae* can be described as truly cosmopolitan with the exception of the polar regions. It occurs in all major ocean basins, in central waters, near oceanic islands, near continental slopes. It occurs in warm, temperate, and sub-boreal waters." A species that is found all over the world, and in such a variety of habitats, is a rarity among cephalopods.

---

*In 1980, three specimens, one "in almost perfect condition," were found floating offshore in South Australian waters. When Wolfgang Zeidler of the South Australian Museum described them, he wrote, "Nearly all known specimens of *Taningia* have been collected from sperm whale stomachs, and it is unusual to encounter them floating at the surface. It is possible that they were regurgitated by sperm whales, and this may be the case for the specimen lacking a head, but the other two were in relatively good condition, and the fishermen estimated that they had died only recently."

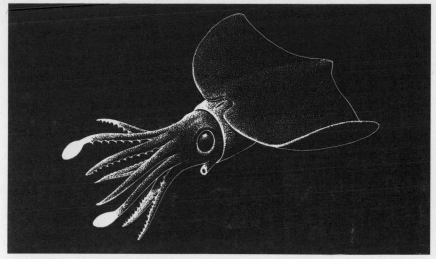

A 7-foot-long squid that has the largest light organs of any animal on the ends of two of its arms, Taningia danae *can flash these lemon-sized photophores like strobe lights.*

Now that hunting of sperm whales has ceased, the modern records must depend on net hauls, which are more useful for determining geographical distribution and other aspects of the biology of *Taningia,* but less useful for collecting the larger specimens. "Our material," wrote Roper and Vecchione, "expands the knowledge of this species through a number of specimens collected in a variety of nets, not predators."

A small specimen (23.4 inches in mantle length), captured at night off Hawaii, was alive and undamaged and placed in a shipboard aquarium, where it was observed and photographed. Although Joubin's 1931 description of the species suggested that the organs at the tips of the arms were photophores, their luminescence had never been directly observed. As discussed in Clyde Roper and Michael Vecchione's 1993 review, Roper and Richard Young "placed the specimen in a cold water aquarium (5°C) in complete darkness, and the observers' eyes were allowed to adapt thoroughly to the dark. Without the aid of a light one observer slowly moved his hand around in the aquarium in an effort to stimulate a response." This is what happened:

> Two primary responses were evoked. Both included bright flashes of brilliant
> blue-green light simultaneously from both arm-tip photophores. The most com-

mon reaction involved the coordinate flashes accompanied by an attack, grasping the researcher's fingers and biting. The second reaction . . . involved a bright flash followed by a rapid retreat from the stimulus. The flashes appeared to vary somewhat in intensity and duration. Usually the flashes lasted only a fraction of a second, but occasionally the organs glowed with fluctuating intensity for 1–7 seconds. These prolonged glows were associated with continuous stimulation, such as pinching of the fins. During these prolonged glows the intensity of the light appeared to increase gradually to a peak and then it receded gradually.

Very interesting for Roper and Young, but what did these responses suggest about the natural behavior of *Taningia*?

The bright, quick arm-tip flashes startled the observers, and created the impression that the flashes serve to startle, distract, and confuse an approaching predator. How effective this is on smaller predators can only be imagined at this time. Clearly large predators are not always foiled, because they are by far the largest source of specimens of *T. danae* in collections. The predators include visual hunters such as tunas and lancetfishes. Perhaps because sperm whales hunt using sonar rather than vision, flashing by *T. danae* is a particularly ineffective defense against these predators.

Even though they do nothing to deter a 60-foot-long sperm whale, the light organs of *Taningia* must be among the most terrifying sights in the blackness of the abyss—if the prey manages to survive the shock of a 7-foot-long carnivorous squid with stroboscopic arm flashers.

Taningia danae is included in the family of eight-armed squids (Octopoteuthidae) because while the juveniles have two tentacles in addition to the eight arms, by the time they mature, the tentacles are reduced to rudimentary filaments or disappear altogether. They are worldwide in distribution, and are believed to live at depths down to three thousand feet. In his 1967 discussion of this species, Malcolm Clarke wrote that "in the smallest individual, chromatophores are dotted over the surface, while the larger specimens have an evenly colored magenta skin." In tabulating the squid species eaten by sperm whales off Japan, Okutani and Satake identified two specimens of *Taningia danae* from whales taken north of Honshu Island. Unlike some of the Atlantic specimens, these two were small, measuring 35.5 cm (13.8 inches) and 47.5 cm (18.5 inches) in dorsal mantle length.

Because sperm whales have been hunted commercially for almost three centuries, we have had more than ample opportunity to examine their stomach contents. In the early days of the fishery, a great deal of potential information was lost. The blubber was stripped off alongside the ship, and the carcass, along with the stomach contents, was discarded. It was only when the whales were hauled up on the decks of the great factory ships that the stomach contents were spilled out on deck. Nevertheless, even before the days of mechanized whaling, whalers observed their quarry, in its death throes, vomiting up great hunks of what could only have been giant squid. As we might expect, Melville discusses this phenomenon in *Moby-Dick:*

> For although other species of whales find their food above water and may
> be seen by man in the act of feeding, the spermaceti whale obtains his food in
> unknown zones below the surface; and only by inference is it that anyone can
> tell of what, precisely, that food consists. At times, when closely pursued, he will
> disgorge what are supposed to be the detached arms of the squid; some of them
> thus exhibited exceeding twenty or thirty feet in length. They fancy that the
> monster to which these arms belonged ordinarily clings by them to the bed of
> the ocean; and that the sperm whale, unlike other species, is supplied with teeth
> in order to attack and tear it.*

And some fifty years later (in *Denizens of the Deep,* a natural history of sea creatures), Frank Bullen wrote much the same thing:

> Every officer, to say nothing of the men, must have known of the very real
> existence of the great Squid, since scarcely a sperm whale can be killed without

---

*For all his genius and historical accuracy, Melville turned out to be wrong about both the squid *and* the whale. Squids—giant or otherwise—do not cling to the seabed with their arms; and the teeth of sperm whales, located in only the lower jaw, are probably used only to capture the squids, pincer-fashion, not to tear them. For the most part, squids that have been examined from the stomachs of sperm whales exhibit no tooth marks or punctures, and it is now assumed that the sperm whale captures its prey by emitting focused sound beams of such intensity that they can stun or even kill the prey.

first ejecting from his stomach huge fragments of this popularly believed by seamen to be the largest of all God's creatures. Not only so, but in every book which has been written about the sperm whale fishery some allusion to the great Cuttle-fish will surely be found, although it must be admitted that so much superstitiously childish matter is usually mixed up with the facts as to make the latter difficult of belief.

In a fictional digression, aboard a ship he calls the *"Cachalot,"* Bullen espied "great masses of white, semi-transparent-looking substance floating about, of huge size and irregular shape," and asked the mate to tell him what they could be. "When dying," the mate explained, "the cachalot always ejected the contents of his stomach, which were invariably composed of such masses as we saw before us; he believed the stuff to be portions of a big Cuttle-fish, bitten off by the whale for the purpose of swallowing." Bullen hooked one of the lumps, and drew it alongside:

> It was at once evident that it was a massive fragment of Cuttle-fish— tentacle or arm—as thick as a stout man's body, and with six or seven sucking discs or *acetabula* on it. These were about as large as a saucer, and on their inner edge were thickly set with hooks and claws all round the rim, sharp as needles, and almost the size and shape of a tiger's.

In *Denizens of the Deep*, Bullen speculated about the relationship between the sperm whale and the giant squid:

> The gigantic Cuttle-fish must be very prolific. He is the principal food, the main support of the sperm whale, and as this vast mammal's numbers are incalculable, and each individual needs, at the very lowest computation, a ton of food to keep him going, the numbers of mollusca upon which he feeds must be proportionate. As to the numbers of sperm whales I may say in passing, that it has several times been my lot to witness an assemblage of cachalots, all of the largest size, covering an area of ocean as far as the eye could reach from the masthead of our ship in every direction. . . . Only to think of the amount of food required for that stupendous host makes my mind reel.

While Bullen's mind reeled, others tried to find out how much food a sperm whale actually needed. In his comprehensive study of the whales captured from

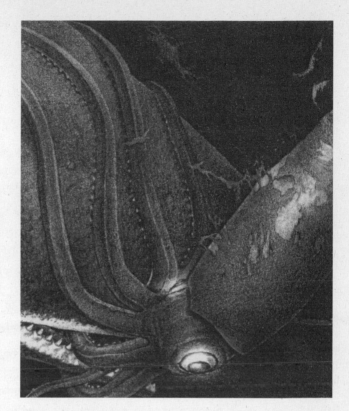

Wrapped in the arms of a giant squid, a sperm whale dives for the depths. The squid is probably better adapted to continue the battle underwater, since, unlike the whale, it does not have to surface for air.

the Durban fishery between 1926 and 1931, L. Harrison Matthews examined the stomachs of eighty-one sperm whales, and of these, the stomachs of nearly all contained the remains of cephalopods, among other things. Most of them were small, averaging about 3 feet in length, but the "very large cephalopods were represented only by beaks in the stomachs and scars on the skin." There is no question, however, that sperm whales occasionally battle and ingest giant squid: according to Rees and Maul, one was even regurgitated in a state where it still showed signs of life. A whale harpooned off Madeira in 1952 had vomited up a 34-footer that weighed about 330 pounds and writhed on the flensing deck until it expired. In 1955, Robert Clarke was present at the whaling station at Porto Pim on the Azorean island of Fayal when a giant squid was discovered in the stomach of a 47-foot-long whale. It weighed 405 pounds, and measured 34 feet, 5 inches from the tip of the tail to the tip of the longest tentacle.

The digestive juices of sperm whales are strong, and the remains of their food items are often corroded beyond recognition. Squid beaks, however, are composed of a tough, chitinous material, and resist digestion much more successfully than the soft parts.* A branch of teuthology, emphasized and practiced by Malcolm Clarke, involves the use of squid beaks as a way of identifying the species (it is fairly easy to identify the beak of *Architeuthis;* it can be 6 inches long). That sperm whales do indeed capture and eat giant squid can be seen in Fiscus and Rice's examination of the stomach contents of sperm whales that were collected off the coast of California from 1959 to 1970: 12 of the 552 whales examined had mandibles of giant squid in their stomachs. But, wrote the authors, "sperm whales may eat *Architeuthis* more often than our records indicate. . . . *Architeuthis* mandibles could be overlooked among remains of *Moroteuthis robustus,* another 'giant' species (although smaller than *Architeuthis*) that is the predominant food of sperm whales off California."

But how do sperm whales catch large squids? Unlike many other predator-prey interactions, this one is not at all self-evident. Lions and zebras both live on the African plains. Lion sees zebra, stalks same. Lion chases zebra, catches zebra, feeds on zebra. Both the predator and the prey live in the same place, and breathe the same stuff in more or less the same way. Now consider bats. They live on insects that they catch on the wing, but the rules are somewhat different. Lions can see the herds of zebras, but the bats have to be able to find and capture the insects in the dark. To accomplish this, bats have developed a highly sophisticated sonar system in which they emit high-pitched sounds that bounce off flying insects; the bats then read the returning echoes and adjust their flight to intercept and capture the insects. Although this is a far more complex system than the lion-zebra relationship, it is similar in that both creatures are capable of flying through the air at night. And again, they breathe the same stuff, so neither one is placed at a particular disadvantage by the medium in which they live.

The sperm whale, like the lion and the bat, is a mammal and breathes air; its

---

*Although the actual process of its formation is unknown, the material known as ambergris occurs only in the intestinal tract of sperm whales, and can be found either in the whale or vomited up and floating on the surface of the ocean. It is a grayish, crumbly material, often compared to peat moss, that somehow forms around a squid beak. In the past it was worth more than its weight in gold, and was used as a fixative for perfumes. The largest lump ever recorded weighed 983 pounds.

prey, however, does not. But, you will say, seals, sea lions, and otters are also mammals, and they feed on fish that get their oxygen without coming to the surface. Seals, sea lions, and otters chase their prey visually, and while some kinds of seals—Weddell seals, for example—are prodigious divers, most of them hunt where they can see what they are chasing. (Some penguins are also unexpectedly deep divers.) Sperm whales, however, hunt in a manner that we do not understand, because it has always been hidden from the eyes of investigators by the inaccessibility and opaque darkness of the environment in which it occurs.

Sperm whales take oxygen into their lungs just they way you do: they inhale a breath of fresh air. (Their "spouts" are their exhalations.) But the hunted cephalopods obtain their oxygen from water, and therefore do not have to surface to breathe. Herein lies the greatest conundrum of the sperm whale's hunting: how can it find and capture enough food on these deep dives, especially if the prey is a more agile and faster swimmer and, more critically, if the whale has to abandon the hunt after a certain period of time because it is running out of air?

Because they write about it extensively, we know how humans catch squids,* but for the whales, how the squids are transformed from fast, free-swimming animals into stomach contents is still a mystery. Like all odontocetes (toothed whales), sperm whales are known to echolocate. From an apparatus in the head, sperm whales (and most dolphins) can broadcast high-frequency sounds directionally into the water, and then read the returning echoes for information on the identity (and perhaps such things as the condition, speed, and texture) of the object. The echolocation of dolphins has been long recognized, but the mechanics whereby the animals actually *catch* their prey was more problematical. It is one thing to locate, say, a school of small squid, but quite another to catch enough of

---

*Commercial squid boats employ lights—underwater and topside—to attract the squids, and "jigs" to catch them. Until about 1950, hand lines were used, but then the Japanese, the leaders in squid-fishing technology, developed automated jigging machines, which deploy numerous jigs concurrently. The jig itself is a spindle-shaped lure with two or three rows of barbless hooks in a ring around the end away from the eye, with which it is fastened to the line. Either by hand or by machine, the lures are "jiggled" in the water to attract the squids. Most of the technology was developed in traditional fisheries for the common Japanese squid *(Todarodes pacificus)*, but the equipment has been modified for catching the much larger and more powerful *Dosidicus gigas* (Hamabe *et al.,* 1982).

them to make a meal. After all, the squid don't have to surface to breathe, but the whales have to do their food locating and food catching while holding their breath, often in the darkness of the depths.

Even when the echolocating capabilities of odontocetes were understood, there was a piece missing from the puzzle, since the first cetological acousticians simply assumed that the whales found the squids by listening to their echoes, and then dashed around gobbling them up. Upon reflection, this did not appear to be a terribly efficient method of hunting, especially considering the speed and maneuverability of the prey, and also its inherent unwillingness to be eaten. The sperm whale has massive peglike teeth in its lower jaw, so if the whale chased down its prey and snagged it in its jaws, the squids ought to have shown some evidence of having been bitten, but they didn't. One of the first to notice this was the Soviet cetologist A. A. Berzin, who published an exhaustive study of the sperm whale in 1971. After examining the stomach contents of a large number of sperm whales, he wrote:

> The mystery of how this whale feeds deepened in view of the following circumstances. . . . Beale gives an example of a capture of sperm whales in normal condition, one of which was blind while two others had deformed jaws. Up to 10 sperms with badly deformed jaws were recorded in our materials. They were in the same condition as all the other animals and had well-filled stomachs, the contents of which did not differ qualitatively from those of other sperm whales caught the same day. . . . All the above suggests that neither the teeth nor the lower jaw need to participate in obtaining food and in the digestive process.

If the sperm whale does not use its jaws and teeth to capture its food, what does it use? In his study, Berzin reviewed the earlier theories, one of which had been propounded by Thomas Beale, a British surgeon who shipped aboard the whaler *Kent* in 1831–32. Upon the conclusion of the voyage, he wrote *A Few Observations on the Natural History of the Sperm Whale,* which was published in 1835, and served as one of the primary sources for the cetology chapters in *Moby-Dick.*

In this work, Beale mentioned several large cephalopods, including the one "discovered by Drs. Banks and Solander, in Captain Cook's first voyage [that] must have measured at least six feet from the end of the tail to the end of the tentacles. . . . But this last," he wrote, "we must imagine a mere pigmy, when we consider the enormous dimensions of the one spoken of by Dr. Schewediawer, in the Phil. Trans. . . . whose tentaculum or limb measured twenty-seven feet in length." The good doctor's remarks, which were actually presented to the Royal Society by Sir Joseph Banks, include, as a sort of a footnote, a story that he heard from someone else, for he says, "One of the gentlemen who was so kind as to communicate to me his observations on this subject, about ten years ago hooked a spermaceti-whale that had in its mouth . . . a large dentaculum of the Sepia Octopodia nearly 27 feet long. This dentaculum did not seem to be entire, one end of it appearing in some measure corroded by digestion, so that, in its natural state it may have been a great deal longer."*

Like most sperm whalers, Beale could not imagine how "such a large and unwieldy animal as this whale could ever catch a sufficient quantity of such small animals, if he had to pursue them individually for his food," and suggested that the whale descends to a certain depth, where it "remains in as quiet a state as possible, opening his narrow elongated mouth until the lower jaw hangs down perpendicularly." The jaws and teeth, wrote Beale, "being of a bright glistening white colour . . . seems to be the incitement by which the prey are attracted, and when a sufficient number are in his mouth, he rapidly closes the jaw and swallows the contents."

In his extensive study called *Whales,* the Dutch cetologist E. J. Slijper suggested a modification of Beale's theory. He wrote: "It is believed that sperm whales do not so much go after this prey as swim about with open mouths, enticing the cuttlefish which seems unable to resist the colorful contrast between the sperm whale's purple tongue and the white gum of the jaws." Slijper knew more about whales than he did about squids, because in the blackness of the depths, the cuttlefishes could not see the whale's tongue, its gums, or any other part of it. Berzin noted that "at a depth of some 100 meters, both prey and hunter are invisible to each other. Moreover, squid are known to be much more mobile than sperm

---

*Beale's quote from the *Philosophical Transactions of the Royal Society of London* for 1783 does not exactly reproduce Dr. Schwediawer's words. Beale called him "Dr. Schewediawer," and referred to the "dentaculum" as a "tentaculum." For the most part, however, his version is correct; I have taken the liberty of replacing his transcription with the actual text of the original.

whales . . . even small squid can develop speeds of up to 40 km an hour, and larger specimens still higher speeds, far outdistancing sperms." Obviously, there had to be some device whereby the whale could stop or at least slow down the squids to capture them, and indeed there was. Even earlier than Berzin, another pair of Soviet cetologists, Vladimir Bel'kovich and Alexei Yablokov, had calculated the sound intensity that might be developed in the sperm whale's nose, and in 1963 they collaborated on a short article entitled "The Whale—an Ultrasonic Projector," in which they suggested that the whale might somehow use its nose to project sounds loud enough to stun its prey. Berzin wrote that by concentrating its sound beam on a selected object, "the animal can create a short-term pressure which must act as an ultrasonic blow capable, even if briefly, of halving, stunning, and paralyzing the object."*

In 1983, Kenneth S. Norris, of the University of California at Santa Cruz, and Bertel Møhl, of Aarhus University in Denmark, published their hypothesis that odontocetes could indeed debilitate their prey with sound. Other theories, such as Beale's, above, or Clarke's, where the whale maintains a position of neutral buoyancy and waits for a school of squid to swim within range, did not explain the uninjured state of the squids in the sperm whale's stomach, nor did these other theories explain how a large animal like the sperm whale could obtain the twenty-two hundred to forty-four hundred pounds of food per day that would be required to sustain it. The "sonic boom" hypothesis not only explains how the cumbersome sperm whale hunts and captures the swift cephalopods, but it also answers many other questions that had heretofore been as problematical as the feeding technique.

Another theory about how sperm whales locate their food suggests that the whales might provide their own light. If the whales' sound bursts cause deep-water dinoflagellates to light up, their bioluminescence might provide enough light for the hunting whale to see—and therefore catch—its prey in what would

---

*According to V. A. Kozak, another Soviet cetologist, in order to be able to hunt in complete darkness, the sperm whale has developed "a unique video-receptor system in the process of evolutionary transformation, which lets the animal obtain the image of objects in the acoustic flow of reflected energy even in complete darkness." In other words, the sperm whale has a sort of audiovisual system that transforms sound into "images," employing the "blisters" on the rear wall of the whale's nasofrontal sac, located in the hollowed-out cradle of the skull.

As depicted by wildlife artist Francis Lee Jaques, a sperm whale pursues a school of (non-giant) squid.

otherwise be total darkness.* This does not explain the lack of tooth marks on the captured squids, nor does it account for the squid's ability to outswim and out-maneuver a slow swimmer like a sperm whale. This startling theory appeared without attribution in the National Geographic Society's 1995 *Whales, Dolphins and Porpoises* (by Darling, Nicklin, Norris, Whitehead, and Würsig). Because *Architeuthis* doesn't light up, this idea, however far-fetched, might be used to explain how sperm whales can find the great squid in the depths.

If the sperm whale debilitated or killed its prey with sound, that would go a long way to explaining the unusual construction of the jaws; the whale could use its tooth-studded lower jaw pincer-fashion to pluck the floating squid out of the water or off the bottom—which would then account for the lack of tooth marks

---

*Even less is known about how squids catch their prey in the depths, but a study published by Fleisher and Case in 1995 shows that some squids hunt by the light generated by their stimulation of light-emitting zooplankton. In the laboratory, *Sepia officinalis* (the common cuttle-fish) and *Euprymna scolopes* (a small squid from Hawaiian waters) were observed to capture nonbioluminescent prey much more frequently and efficiently when the water was illuminated by the movement-stimulated *Pyrocistis fusiformis,* a bioluminescent dinoflagellate. If the prey does not emit light, the predator might have to provide it.

on the prey. If the squid floated to the bottom, the sperm whale might plow its lower jaw through the sediment to pick the squid up, which would explain the strange items occasionally found in sperm whale stomachs. In their study entitled "Stones and Other Aliens in the Stomachs of Sperm Whales in the Bering Sea," Takahisa Nemoto and Keiji Nasu identified stones, sand, crabs, glass buoys, a coconut, and a deep-sea sponge as coming from the stomachs of sperm whales killed by Japanese whalers in the vicinity of the Aleutian Islands. (The largest stone they found weighed just over three pounds.) It would also account for the occasional entrapment and drowning of sperm whales in undersea telegraph cables: plowing along the bottom, the whale might accidentally become entangled in a loop of cable; or it might even have mistaken the cable for the tentacle of a giant squid. In a 1957 study entitled "Whales Entangled in Deep-Sea Cables," oceanographer Bruce Heezen listed fourteen instances of sperm whales trapped and drowned in cables, and wrote, "It is possible that the whales attack tangled masses of slack cable mistaking them for items of food." The deepest recorded entanglement was 620 fathoms, or 3,720 feet.

From the examination of the stomach contents spilled on the decks of whaling ships, we know that the whales eat lots of squids, and based on the number of squid beaks found in the stomachs of captured sperm whales, we can get an idea of the quantity. Finds of five thousand to seven thousand beaks per whale are not uncommon, and Berzin mentions one Soviet scientist who found twenty-eight thousand beaks in the stomach of a single whale, indicating a feeding frenzy in which fourteen thousand squids were consumed. (Each beak has a separate upper and lower mandible.) It would appear that such consumption would require a dense concentration of squids, and indeed, squids may be the most numerous large animals in the ocean.

Although his evidence is largely circumstantial (or nonexistent), Ivan Sanderson (in *Follow the Whale*) discusses the numbers of squids required to feed the world's population of sperm whales:

> Most people don't even know what a squid is, yet these animals probably make up a greater aggregate of pure animal matter on this earth than any other two kinds of living creatures put together. They exist in countless millions of apparently endless masses in every ocean and sea in the world, and almost three-quarters of this

planet is covered by oceans and seas which are on the average nearly two and a half miles deep. Throughout this vast volume of liquid there are probably more squids than anything else.

Malcolm Clarke, a British scientist who specializes in sperm whales and squids, commented on the complex interaction of the two in a 1977 study:

> Man's awareness of the existence of large squid came, not from what he caught in his nets, but from monsters floating dead or moribund at the sea surface and from the tales of whalers who had seen, with unbelieving eyes, whales vomit complete or dismembered kraken of immense proportions. Such doubtful tales hardened into drawings and recorded measurements over one century ago, and ever since, man has tried desperately to catch by net and line, these will-o'-the-wisps of the sea. Though our nets have become larger and larger and faster and faster, very little progress and most of that in the last decade, has been made towards catching any deep sea squids greater than half a meter or so in length. In a century, many tantalising glimpses of the deep sea squids have come from strandings on the coast and from the stomachs of toothed whales, particularly the commercially exploited Sperm Whale.

In this study, Clarke estimated the amount of food required to feed the enormous world population of sperm whales. (Recognizing the difficulties of estimating whale populations, he wrote, "Estimates of the whale population are, unhappily, notoriously questionable, but a 1973 estimate placed this at 1¼ million.") Using a mean weight of 15 tons for males and 5 tons for females, Clarke arrived at a total weight of the world's sperm whales of 10 million tons, which would require *100 million tons of squids per year.* (My astonished italics.) This is larger than the biomass of the annual world catch of fish by fishermen "and probably approaches the total biomass of mankind." In other words, the weight of squids eaten every year by sperm whales is greater than the weight of the entire human race.

Interviewed by Richard Conniff for an 1996 article in *Smithsonian* magazine, Clyde Roper said, "I've looked at the biology of the giant squid, and also of the sperm whale, the records of where strandings occur and the best concentrations of squid in stomach contents of sperm whales." He said he interviewed whalers in the Azores who used to harpoon sperm whales by hand from small boats. When they were harpooned, the whales vomited up *lula grande* (giant squid). Roper asked how often they ran into this, and they said that virtually every whale would have a giant squid in it. "You do the arithmetic—there are perhaps a million sperm whales in the world—and it follows that there have to be a lot of giant squid around."*

Because the biology of squids is so poorly known, one of the best tools for learning about them is the examination of sperm whale stomach contents. We can therefore not only ascertain the numbers of squids consumed, but we can also learn a great deal about the squids themselves. Indeed, the whales perform an important service for those who would study squids; no other way of collecting is nearly as productive. Squids can often evade nets and trawls, but they seem to be less successful at avoiding the powerful, deep-diving whale that feeds on them. (One of the disadvantages of this method, however, is that the specimens are often partially digested by the time the teuthologists get to look at them.) One of the most useful books ever written about squids is Malcolm Clarke's massive *Cephalopoda in the Diet of Sperm Whales of the Southern Hemisphere and Their Bearing on Sperm Whale Biology,* published as a *Discovery Report* in 1980. As

---

*Alan Pampanin, Bennett Savitz, and Michael Greenberg, who identified themselves as members of the "Calamari Legal Institute" of Cambridge, Massachusetts, did the arithmetic as Roper suggested, and this is what they wrote in a letter to the editor of *Smithsonian:* "In his May 1996 article on the giant squid, Richard Conniff reported two hard-to-fathom findings by scientist Clyde Roper. First Roper estimated that a 50-ton sperm whale may eat three or four giant squid per day. Second, Roper said that there are roughly a million sperm whales in the world. Conservatively estimating, if only half of the whales are eating three squid per day, a cataclysmic calamari ingestion of 547,500,000 giant squid is occurring annually. Extrapolating further, if one assumes that the volume of a giant squid is roughly 400 cubic feet, a gargantuan 219 billion cubic feet of giant squid are being eaten by voracious sperm whales on a yearly basis. This figure, which comes to a voluminous 4,211,538.461 cubic feet of squid being consumed per week, does not even include the millions of giant squid that must *not* be eaten on a weekly basis in order for them to reproduce for those deep-diving dirigibles of death, the sperm whales.

"The moon, as it turns out, is a little more than five billion cubic feet. We estimate then, that if the moon were made of giant squid (as opposed to green cheese), about 80 percent of it would be consumed in one week by the sperm whales of this planet. We at the Calamari Legal Institute are having trouble digesting these figures."

The following month, in response to *their* letter, Gary Garrett wrote: "I may have some digestive relief for the Calamari Legal Institute. The volume of the Moon—whether it's made of squid or green cheese—is 5 billion cubic *miles,* not feet. Insatiable sperm whales could not consume a calamari Moon, therefore, for billions of years."

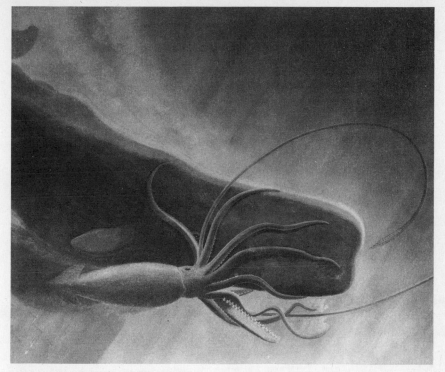

Giant Squid in Battle with Sperm Whale, *by Francis Lee Jaques. This grisaille (black-and-white) painting was used as a basis for the diorama in the American Museum of Natural History.*

Clarke wrote in the introduction to this 324-page study, "Whale and squid biology are clearly closely linked, and consideration of both subjects in one paper is necessary to avoid duplication." Since beginning the study in 1962, Clarke (and various colleagues) have examined the stomach contents of 461 sperm whales, collected at the whaling stations of Durban and Donkergat in South Africa; Cheynes Beach in Albany, Western Australia; the island of South Georgia; and the British pelagic factory ships *Southern Harvester* and *Southern Venturer.*

Of our titular subject, he said, "Twenty-three buccal masses and eight large pieces of flesh belonging to *Architeuthis* were collected from whales caught off Durban, Donkergat and Albany." Despite the popular perception that sperm whales feed regularly on giant squid (see Roper's statement above), Clarke's research showed that "*Architeuthis* only comprises 0.26% of the beaks collected." The largest *Architeuthis* beak represented a squid weighing 120 to 180 kg (264 to

396 pounds), probably not in the "monster" class. There were no stomachs sampled with the remains of more than one *Architeuthis,* which suggested that the whale was catching solitary animals, or that "their large size may facilitate escape if a shoal* is attacked."

One thing that we cannot learn from the examination of sperm whales' stomach contents is where in the water column *Architeuthis* lives or hunts. (Other species of squid have been caught in nets and trawls at certain depths; Roper and Young produced an extensive study entitled "Vertical Distribution of Pelagic Cephalopods" in 1975, but of *Architeuthis,* they wrote, "Very little is known of its vertical distribution or any other aspect of its biology.") Sperm whales are among the deep-diving champions of the mammalian world, able to dive to known depths of ten thousand feet and to hold their breath for an hour and a half. They would therefore be able to capture giant squid at any level that they can reach. Two juvenile giant squid were obtained from fish stomachs, but most of what we know about the distribution of giant squid—vertical or otherwise— comes from the squid appearing on the beach, at the surface, or spilled out of a whale's stomach. The stomach contents are obviously related to where the whalers caught their prey, and just as the whalers hunted where the whales were supposed to be, so might the whales have gone to those locations where the squid were supposed to be.

Sperm whales have the largest brains of any animals that have ever lived. They lead complex social lives and are capable of making a great variety of sounds that are undoubtedly part of an intricate communications system. The immediate future holds little promise of interspecies communication, but if any creature can reveal the secret life of *Architeuthis,* it would certainly be *Physeter macrocephalus,* the great sperm whale, which knows enough about the giant squid to find it in the darkness of an icy ocean, a mile below the surface.

---

*In British ichthyological terminology, the word *shoal* is used where Americans use *school.* The words come from the same sixteenth-century Dutch root, *schole,* which referred to a group of aquatic animals or floating objects, and has nothing to do with education.

The Giant
Squid in
Literature
and Cinema

There is probably no apparition more terrifying than a gigantic, saucer-eyed creature of the depths with writhing, snakelike, grasping tentacles, a huge gelatinous body, and the powerful beak of a humongous seagoing parrot. Even the man-eating shark pales by comparison to such a horror. In only a few species do octopuses and squids reach the monstrous size required to attack swimmers, but this has not deterred writers and filmmakers from recruiting them where an underwater man-killer is needed.

An illustration from the original edition of Twenty Thousand Leagues Under the Sea. The illustrator, identified only as "de Neuville," drew the animal as a gigantic squid.

Probably the best-known scene in *Twenty Thousand Leagues Under the Sea* is the attack on the submarine *Nautilus* by a giant squid. (If it is not the most celebrated episode in the book, it is certainly the scene most people remember from the 1954 movie.) In this scene, after a characteristic discussion of the immediate perils that might await them (and some historical references as well), Aronnax, Conseil, and Ned Land espy "a terrible monster worthy of all the legends about such creatures:"

It was a giant squid twenty-five feet long. It was heading toward the *Nautilus*, swimming backward very fast. Its huge immobile eyes were of a blue-green color. The eight arms, or rather legs, coming out of its head—it is this which earned it the name of "cephalopod"—were twice as long as its body and were twisting like the hair of a Greek fury. We could clearly make out the 250 suckers lining the inside of its tentacles, some of which fastened onto the glass panel of the lounge. The monster's mouth—a horny beak like that of a parakeet—opened and closed vertically. Its tongue, also made of a hornlike substance and armed with several rows of sharp teeth, would come out and shake what seemed like veritable cutlery. What a whim of nature! A bird's beak in a mollusk! Its elongated body, with a slight swelling in the middle, formed a fleshy mass that must have weighed between forty and fifty thousand pounds. Its color, which could change very fast according to the animal's mood, would vary from a ghastly gray to reddish brown.

The cover of the *Classic Comics* version of Twenty Thousand Leagues Under the Sea. *The malevolent giant squid has a rather unusual funnel arrangement, with one under each of its eyes.*

From the number of arms (or legs) to the eye color, body color, and the nature of the tongue, Verne has almost everything wrong about the giant squid. It is beyond merely "wrong" to attribute a weight of "forty or fifty thousand pounds" to an animal that is 25 feet long and mostly tentacles; to weigh this much it would have to have been made of iron. A 25-foot minke whale, which is all meat and muscle and has no long, skinny arms, weighs about 10 tons. Later, as the squid tries (and fails) to hold onto the submarine with its suckers, Aronnax exclaims, "What vitality the Creator has given them, and what vigor of movement! And to think they possess three hearts!"

In *Twenty Thousand Leagues,* the squid attacks the submarine, presumably to get at the people, and because Nemo's electric bullets would not affect the cephalopod ("their soft flesh doesn't offer enough resistance to make them explode"), the crew takes after it with axes and Ned Land's harpoon. As the submarine surfaces and the hatches are opened, one of the squid's arms snakes down the hatchway, where Nemo lops it off with a blow of his boarding ax. With seven of its eight arms cut off, the monster still manages to capture one of the sailors. The poor wretch is dragged off in a cloud of ink, and the crew attacks the remaining ten or twelve squid that "had invaded the platform and sides of the *Nautilus.*" As brave Ned Land harpoons one of the monsters, he becomes trapped, and as a squid is about to chomp down on him with its huge beak—likened to that of a parakeet (the original says "*comme le bec d'un perroquet*")—Nemo saves him and buries his ax between the two mandibles. Land then plunges his harpoon deep into the creature's triple heart, and the episode draws to a merciful conclusion.

On other occasions when the giant squid appears in literature, it fulfills, or often exceeds, its reputation. When H. G. Wells wrote a story he called "The Sea Raiders" in 1905, it had only been twenty-odd years since a large number of giant squid had washed ashore in Newfoundland. In this story, Wells changes the locale to Sidmouth, on the coast of Devonshire. He invents *Haploteuthis ferox* (which can be translated as "one fierce squid"), and then he has them attacking a Mr. Fison, who has spotted some of them feeding on the carcass of a man. When Fison goes to investigate, he sees that

> the rounded bodies were new and ghastly-looking creatures, in shape somewhat resembling an octopus, and with huge and very long and flexible tentacles, coiled copiously on the ground. The skin had a glistening texture, unpleasant to see, like shiny leather. The downward bend of the tentacle-surrounded mouth,

the curious excrescence at the bend, the tentacles, and the large intelligent eyes, gave the creatures a grotesque suggestion of a face. They were the size of fair-sized swine about the body, and the tentacles seemed to him to be many feet in length. There were, he thinks, seven or eight at least of the creatures.

Fison tries to drive them off by shouting, but when he makes the mistake of throwing a rock at them, they chase him up the beach: "And then, slowly uncoiling their tentacles, they all began moving towards him—creeping at first deliberately, and making a sort of purring sound to each other."

The hapless Fison escapes the amphibious squid by climbing a cliff ("At one point he could hear the creatures splashing in the pools not a dozen feet behind him"), but when the cephalopods return to the water, people in boats do not fare so well. The squid congregate just offshore ("Then these things, growing larger, until at last the bottom was hidden by their inter-coiling forms, and the tips of their tentacles rose darkly here and there into the air above the swell of the waters"), pulling fishermen overboard, and then eating all the occupants of an excursion boat: three women, a child, a boatman, and "a little man in a pink-rib-

The poulpe "brandishes its victim like a feather" as stalwart crewmen of the Nautilus hack away at it. In the lower left, one of the arms is correctly shown as the clublike tentacle of a squid.

boned straw hat." The image of swarms of giant, noisy squid in shallow offshore waters is the stuff of nightmares, and if they should ever come *ashore*. . . . In all, the "raiders" from the depths kill eleven people, and then, as mysteriously as they arrived, they depart from Sidmouth. Except for a couple of carcasses that wash ashore—one "slashed deeply with a cutlass wound"—they are never seen again.

In addition to those tales that he tried to pass off as factual accounts (like *The Cruise of the "Cachalot"*), Frank Bullen also wrote stories that he intended to be read as fiction, collected in a volume called *Deep-Sea Plunderings* (1902). In this book we find a story entitled "The Last Stand of the Decapods," which might be unique in the giant squid canon, since it does not contain a single person getting grabbed or being dragged toward the gnashing beak. It concerns the rivalry between the squid and the sperm whale, and while it has some marvelous imagery, none of it remotely resembles reality.

The tale begins with a lengthy discussion of the natural history of the kraken, with some perfunctory adherence to the facts, but with flights of imagination that would have made H. G. Wells jealous. For example:

> In his proper realm, crouching far below the surface of the sea in some coral cave or labyrinth of rocks, he must present a sight so awful that the imagination recoils before it. . . . He possesses a cylindrical body reaching in the largest spec- imens yet having recorded as having been seen, a length of between sixty and seventy feet, with an average girth of half that amount. That is to say, consider- ably larger than a Pullman railway car. . . . So far, there is nothing particularly striking about the appearance of this mighty cylinder except in colour. This characteristic varies in different individuals, but is always reminiscent of the hues of a very light-coloured leopard; that is to say, the ground of a livid green- ish-white, while the detail is in splashes of lurid red and yellow, with an occa- sional nimbus of pale blue around these deeper markings. . . . Our friend then, has ten arms springing from the crown of his head, of which eight are forty feet in length, and two are seventy to eighty.

Then Bullen's story really takes off: it seems that the mighty krakens ruled the seas until the sperm whales realized how tasty they were and began gobbling them up. "This desultory warfare was waged for long, until, driven by despair to a com- munity of interest unknown before, the Krakens gradually sought one another out with but a single idea—that of combining against the new enemy; for, knowing to what immense size their kind could attain in the remoter fastnesses of the ocean, they could not yet bring themselves to believe that they were to become the helpless prey of the newcomers." A convocation of giant squid is something wondrous to behold, and it gives Bullen the opportunity to write, "Only dimly can we imagine what must have been the appearance of those vast masses of writhing flesh, as through the palely gleaming phosphorescence of the depths they sped backwards in leaps of a hundred fathoms each, their terrible arms, streaming behind like Medusa's hair magnified ten thousand times in size, and with each snaky tress bearing a thousand mouths instead of one." The monsters assemble on the seafloor: "They numbered many thousands, and no one in all their hosts was of lesser magnitude than sixty feet long by thirty in girth of body alone. From that size they increased until some—the acknowledged leaders— discovered themselves like islands, their cylindrical carcasses as huge as that of an ocean liner, and their tentacles capable of overspreading an entire village."*

As the giant decapods huddle on the ocean bottom, one of their number brings them the message that ten thousand sperm whales are on the way, and they release "so dense a cloud of sepia that for many miles around the clear blue of the ocean became turbid, stagnant and foul." Undeterred by the cloudy water and poor visibility, the whales "plunged forward into the abyssal gloom; down, down, withal into that wilderness of waiting devils," and shred the squid into fragments, until the survivors "disentangled themselves from the debris of their late associates and returned with what speed they might to depths and crannies, where they fondly hoped that their enemies could never come. They bore with them the certain knowledge that from henceforth they were no longer lords of the sea, that instead of being, as hitherto, devourers of all things living that crossed the radius of their outspread coils, they were now and for all time to be the prey of a nobler race of creatures, a higher order of being, and that at last they had taken their rightful position as creatures of usefulness in the vast economy of Creation."

---

*Surely this must be the ultimate giant squid description. Even disgraceful exaggerations like Cousteau's 24-inch-diameter suction cups (see page 209) pale next to "squid carcasses as huge as that of an ocean liner." At least Bullen had the decency to admit he was writing fiction.

The fearsome reputation of the giant squid makes it a natural for inclusion in movies, but because of its poorly understood morphology and habits, it has been rather difficult to replicate. In 1942, Cecil B. DeMille directed *Reap the Wild Wind,* a lavish extravaganza of sailing ships in the early nineteenth century. (The cast includes John Wayne, Ray Milland, Susan Hayward, Paulette Goddard, Raymond Massey, and Robert Preston.) In the big underwater denouement, two hard-hat divers (Wayne and Milland), rivals for the hand of Goddard, descend to the depth of "ten fathoms" to investigate a wreck that will either exonerate or implicate Wayne of complicity in the wreck of the *Southern Cross.* Enter the villain, a giant squid that has taken up residence in the wreck.

This particular squid is an orange, sponge-rubber creature that emerges from the fo'c'sle and throws a tentacle around Milland. John Wayne, whose career depends upon his rival not finding certain evidence, watches passively as Milland struggles with the squid, but eventually, his better nature takes over, and he enters the battle. Of course, since Wayne is actually guilty, he manages to free Milland, but he has to go down with the squid. In this film, *Architeuthis* probably fulfilled the public's expectations of what this beast is supposed to do: it lurks in drowned ships, awaiting luckless divers. From the vantage point of hindsight, the squid in *Reap the Wild Wind* is not very accurate, and its habits are ludicrous, but it was such a unique addition to the lexicon of the cinema that the film was awarded an Oscar for special effects.

It took the genius of Walt Disney Studios to bring the giant squid to life in the movies. In the 1954 production of *Twenty Thousand Leagues Under the Sea* the animal is portrayed as an aggressive, deep-water predator that attacks the submarine. (Before the attack, Captain Nemo announces that the *Nautilus* has penetrated "deeper than man has ever gone before.") When the approaching squid is first sighted through a porthole of the *Nautilus,* a crewman shouts, "Giant squid astern, sir!" and the battle is joined. The squid grasps the rudder of the submarine, but is dissuaded by a jolt of electricity. Captain Nemo (played by James Mason) brings the *Nautilus* to the surface—in a howling gale, naturally—and the squid follows. It envelops the submarine in its tentacles, and Nemo arms his men with axes, halberds, flensing knives, and harpoons. As he prepares to lead his men into battle, Nemo announces, "You'll be fighting at close quarters with the most tenacious of all sea beasts. Take care of the tentacles; they'll seize anything within reach and hang on to the death!" In a tangled welter of writhing arms and driving rain, men fall overboard, the tentacles probe the open hatchways of the submarine looking for suitable victims, and Nemo shouts that the only way to kill the beast is by striking it right between the eyes. Before Nemo can administer this coup de grâce, his assailant pulls him under-water with a snakelike tentacle, and draws him toward its ominously clicking jaws. Ned Land (Kirk Douglas), who had been imprisoned below, breaks free in time to climb on deck and harpoon the squid right between the eyes, and then dives in to rescue Captain Nemo.

When it was made, this film was considered a marvel of spectacular special effects, especially the manufacture and deployment of the giant squid. For the battle scenes on deck, a two-ton model squid was designed, requiring sixteen men to operate the electronics, hydraulics, and remote controls, and another fifty in the rafters to handle the wires that moved the individual tentacles. The kapok-bodied squid was the creation of sculptor Chris Mueller and mechanical effects expert Bob Mattey, who would go on to design the white shark in *Jaws*. At first, the battle scene was shot against a placid sea with a red sunset sky, but all the wires showed, so the decision was made to change the weather to a storm and film it again. In his memoirs, the director, Richard Fleischer, described the events that led to the reshoot:

> Lighting the set for day not only exposed every wire that was used to manip-ulate the squid but also showed every fault in its construction, and there were many. The deck of the *Nautilus* looked like a concrete slab. With all the sailors and the huge squid, it should have canted a little, or rocked. It didn't. Then there was the squid itself. The stuff it was made of started to deteriorate. Big hunks of the tentacles would drop off when the sailors were wrestling with it. Sometimes whole tentacles would come off and we'd have to stop to glue them back on again.

For the underwater shots, a two-foot-long model squid was used, along with a correspondingly scaled miniature submarine. (In the scene in which the squid first attacks the *Nautilus,* the model squid was positioned with its tentacles wrapped around the rudder and then pulled off. When the film was reversed, it looked as if the squid was approaching and then grabbing the submarine.) It would appear that the inclusion of a giant squid is a sure road to an Oscar; like *Reap the Wild Wind*, this film won an Academy Award for best special effects.

The tired-looking squid in *Voyage to the Bottom of the Sea* (1961) would win no Oscars. It just sits on the bottom, its movement restricted to feeble tentacle wiggling and occasional eye movement. It doesn't have a very big role in the movie, and actually doesn't affect the course of the story at all, but it is "the bottom of the sea," after all, so we have to have a giant squid. For reasons that are altogether too complicated (and too silly) to detail here, a fancy atomic submarine has to "tap in" to an undersea telephone cable. Admiral Harriman Nelson (Walter Pidgeon) sends Captain Lee Crane (Robert Sterling) to splice something or other onto the cable. Lurking around the coral is a flabby, almost immobile giant squid, which snakes its tentacles through the kelp to grab Captain Crane. His fellow divers come to rescue him, brandishing knives and flares, and somehow cause the squid to emit a cloud of bright red liquid. (Blood? Red ink?) With the squid out of the way and the telephone cable tapped, the sub goes on to save the world—but not before a giant *octopus* grabs the sub, and is dispatched by furious jolts of bright blue electricity.

An artificial and tired-looking giant squid was not much of a threat in the 1961 film Voyage to the Bottom of the Sea.

The first of the James Bond adventure films was *Doctor No,* made in 1962, in which the eponymous megalomaniac captures 007, but he escapes, eventually to blow up the entire Caribbean island that Doctor No has selected as his base for the conquest of the world. In the film, a tank equipped with a flamethrower, leaking radioactivity, and gigantic explosions are certainly intimidating enough, but they do not compare with the malevolent presence in Ian Fleming's novel. In the book (published in 1958), Bond escapes from his cell and has to overcome a succession of obstacles, including electrified gratings, freezing and burning air shafts, and a roomful of tarantulas, but when he finds himself trapped in an enclosed deep-water inlet, he realizes that Doctor No has arranged for the ultimate killer to ambush him:

The jacket of the 1986 reprint edition of Ian Fleming's Doctor No. In the novel (but not in the movie), James Bond (007) has to escape from a frightening giant squid that is described as having "two long seizing tentacles and ten holding ones." (In this illustration, nine of these arms are visible, which is one more than the giant squid actually has.)

Below him the water quivered. Something was stirring in the depths, something huge. A great length of luminescent greyness showed, poised far down in the darkness. Something snaked up from it, a whiplash as thick as Bond's arm. The tip of the thong was swollen to a narrow oval, with regular budlike markings. It swirled through the water where the fish had been and was withdrawn. Now there was nothing but the huge grey shadow. What was it doing? Was it . . . ? Was it tasting the blood?

As if in answer, two eyes as big as footballs slowly swam up and into Bond's vision. They stopped, twenty feet below his own, and stared up through the quiet water at his face. . . .

Bond stared down, half hypnotized, into the wavering pools of eyes far below. So this was the giant squid, the mythical kraken that could pull ships beneath the waves, the fifty-foot-long monster that battled with whales, that weighed a ton or more. What else did he know about them? That they had two long seizing tentacles and ten holding ones. That they had a huge blunt beak beneath eyes that were the only fishes' eyes that worked on the camera principle, like a man's. That their brains were efficient, that they could shoot backwards through the water at thirty knots, by jet-propulsion. That explosive harpoons burst in their jellied mantle without damaging them. . . . But the bulging black and white targets of the eyes were rising up towards him. The surface of the water shivered. Now Bond could see the forest of tentacles that flowered out of the face of the thing. They were weaving in front of the eyes like a bunch of thick snakes. Bond could see the dots of the suckers on their undersides. Behind the head, the great flap of the mantle softly opened and closed, and behind that the jellied sheen of the body disappeared into the depths. God, the thing was as big as a railway engine!

As Bond clings to the wire fence that keeps the squid captive, the locomotive-sized cephalopod grabs for him with one of its tentacles, "delicately, like the questing trunk of an elephant," and "like a huge slimy caterpillar, the tentacle walked slowly on up his leg." He can feel the suction cups on his ribs, tearing his flesh and pulling him downward. The squid appears above the surface: "The eyes were glaring up at him, redly, venomously, and the forest of feeding arms was at his feet and legs, tearing the cotton fabric away and flailing back. . . . Now the eyes and the great triangular beak were right out of the water and the beak was reaching up for his feet." But even a giant squid is no match for 007. Just as he is about to be dragged into the "great triangular beak," Bond pulls a spear from his trouser leg and hurls himself at the beast, stabbing it in one of its football-sized eyes. It releases a great cloud of sticky black ink, but it sinks out of sight, leaving Bond free to smother the evil Doctor No by dumping a load of guano on him. For a James Bond film, *Doctor No* was quite unpretentious, and had few of the elaborate special effects that characterized the later endeavors. (Even the detonation of an entire island is accomplished with only modest pyrotechnics.) Obviously, the creation of a believable giant squid was far beyond the intentions (or the capabilities) of the makers of this film.

Regardless of the "facts," authors who require a gigantic, threatening, underwater monster often enlist *Architeuthis*. Michael Crichton (who would go on to write *Jurassic Park*) made the giant squid the archvillain of a science-fiction novel (now a movie) he called *Sphere* for a huge, seamless metal ball found on the floor of the Pacific, half-buried a thousand feet below the surface. This is an extraordinarily complicated book, having to do with space and time travel, black holes, mind control by alien forces, and the power of the human brain to imagine frightening creatures, and by so doing, bring them to life. (It is *Twenty Thousand Leagues* meets *2001* meets *Alien*.) One of the characters in the novel says that he was terrified of the squid in the film of *Twenty Thousand Leagues Under the Sea*, so a real one promptly appears. Crichton's giant squid is not exactly real, but somehow, it is real enough to actually kill people.

Its first victim is "almost entirely crushed," and the skin "looks as if a rough file had gone over it." Collecting the body outside their submersible habitat, the divers are told, "You're not alone out there, and whatever's with you is very damn big." Indeed it is. While its dimensions are never actually given, this specimen is huge (on one occasion it is described as "as big as a house"), and on several occasions, it grabs and shakes the habitat, even sending one of its tentacles inside to feel around for victims—à la the Disney film of *Twenty Thousand Leagues*, which was made some twenty-eight years before Crichton wrote *Sphere*. And while Crichton puts an accurate description of the giant squid into the mouth of one of his characters ("There are at least three distinct species . . . the beak is mounted in a ring of muscle so it can twist . . . and the radula—the tongue of the squid—has a raspy, file-like surface"), he evidently decides that a 1,000-pound animal with 40-foot-long arms isn't scary enough, so he makes it glow green in the dark, and he gives it the capability of flashing its green color on and off. (Because they have no photophores, giant squid do not luminesce.) The divers eventually kill it by running two million volts through the skin of their submersible—another borrowing from the Disney film.

Regardless of how little we know about it, the frightening appearance of the kraken has encouraged people throughout the ages, from Pliny the Elder to Peter the Benchley, to speculate on its nature, often in exaggerated, terrifying, and repulsive terms. For example, here is Frank Bullen, introducing the giant squid in a compilation of fact and fantasy he entitled *Denizens of the Deep*:

> He does not pursue his prey; he waits like some unimaginable spider in the centre of his web of far reaching tentacles, with his huge eyes piercing the surrounding sepia-stained waters until a quiver from one of the outlying arms sets the abyssal mouth agape, the mighty parrot-like mandibles clashing as the struggling victim is conveyed inwards.

In 1992, Arthur C. Clarke wrote an article for *Omni* magazine entitled "Squid! A Noble Creature Defended." Clarke doesn't so much defend the squid as discuss his literary affiliations with it, and he opens the article by announcing that he has just heard Peter Benchley promoting *Beast* on the radio. He then comments, "Good for you, Peter—but why did it take you so long? After all, it's a pretty obvious idea. *I* should know." Evidently, the producers of *Jaws* had asked Clarke to write the screenplay for *Jaws 2*, but he countered with an outline for a story in which *Architeuthis* would be the protagonist. He based this proposal on a short story he had written in 1962, called "The Shining Ones." In the *Omni* article, he wrote, "I couldn't resist calling it [the screenplay] *Tentacles*, although I realized that this would provoke lewd sniggers at the box office."*

In his 1966 *Challenge of the Deep*, Clarke identifies the moment he learned of the giant squid:

> When I was a boy, I came across a picture that has haunted me all my life. It was an illustration in a book about whaling called *The Cruise of the Cachalot* by Frank Bullen, and it showed two incredible sea monsters locked together in a death struggle. One was the great sperm whale—itself a weird enough beast with its square box of a head and the narrow jaw hinged like a saw underneath

---

*Although Clarke had nothing to do with it, there actually *was* a movie called *Tentacles;* and it was so bad that it indeed inspired "lewd sniggers" at the box office. It was an Italian film made in 1977, starring Henry Fonda, Shelley Winters, and John Huston, about a man-eating octopus that gobbles up hapless swimmers and sailors because it has become irritated by "illegal" high-frequency sounds; it is eventually dispatched by a school of trained killer whales.

it. But the thing it was fighting might have come straight from a nightmare. It was a huge, flabby mass sprouting a forest of sucker-studded arms which had wrapped itself around the head and jaws of the whale. From the midst of this tangle of tentacles, two immense eyes stared out with a cold and evil intelligence. It was the most terrifying thing I had ever seen, and I could hardly believe that it was real.

Clarke managed to incorporate this terrifying creature into his 1953 novel, *Childhood's End,* a story of alien "Overlords" that oversee the end of the world, a subject that one might assume has very little to do with giant squid. But even

The illustration from the 1910 edition of Frank Bullen's The Cruise of the "Cachalot" that Arthur C. Clarke remembered from his childhood, and which he called "the most terrifying thing I had ever seen."

here we can see his fascination with this creature, as he has two of the characters prepare a tableau for the "museum" of the Overlords. Although it sounds as if it ought to be a description of a real undersea encounter, it is actually a description of a diorama:

> The long, saw-toothed lower jaw of the whale was gaping wide, preparing to fasten upon its prey. The creature's head was almost concealed beneath the writhing network of white, pulpy arms with which the giant squid was fighting desperately for life. Livid sucker-marks, twenty centimeters or more in diameter, had mottled the whale's skin where those arms had fastened. One tentacle

was already a truncated stump, and there could be no doubt as to the ultimate outcome of the battle. When the two greatest beasts on earth engaged in combat, the whale was always the winner. For all the vast strength of its forest of tentacles, the squid's only hope lay in escaping before that patiently grinding jaw had sawn it to pieces. Its great expressionless eyes, half a meter across, stared at its destroyer—though, in all probability, neither creature could see the other in the darkness of the abyss.

As shown in the illustration from Bullen's book on page 181, Clarke's recollections were faithful, and his description of the "diorama" is remarkably similar to the picture that so haunted him as a child.

Clarke continued his literary love affair with *Architeuthis* in "The Shining Ones," a remarkable story containing technological and teuthological elements that would not actually be discovered for many years to come. Briefly, it concerns an underwater engineering consultant who is summoned by the Russians to repair a mammoth hydrothermal generator that has somehow been damaged at a depth of five hundred fathoms off Sri Lanka. The engineer, a Swiss named Klaus Muller, descends in his minisub, and sees that a great chunk of the heating element has been ripped away. He repairs the damage, surfaces, and reports that he can think of nothing that could possibly have wreaked such havoc with the grid, which resembles a gigantic automobile radiator. On his next dive, however, he spots two 20-foot giant squid, obviously too small to have damaged the grid. As he watches them, however, he realizes that they are flashing their photophores in recognizable patterns. First he sees a flashing pattern that resembles his submarine, and then they create patterns that look like squid. He grasps that (Clarke's italics) "*the squids were talking to each other.*" (Giant squid do not have photophores, and therefore cannot flash messages to each other with lights. But all squids—*Architeuthis* included—can change color rapidly, and it is now believed that these color changes are employed, among other functions, for purposes of communication.) They then flash a "picture" that Muller recognizes as an enormous squid, and he says, "My God! They feel they can't handle me. They've gone to fetch Big Brother." The story is written as a transcript of a tape recording, and it ends with Muller's words: "Joe! You were right about Melville! The thing is absolutely gigan—"

The quote from Herman Melville that Joe alluded to in "The Shining Ones" is as follows: "A vast pulpy mass, furlongs in length and breadth, of a glancing cream-color, lay floating in the water, innumerable long arms radiating from its centre, curling and twisting like a nest of anacondas, as if blindly to clutch at any hapless object within reach," and appears again in Clarke's story called "Big Game Hunt." In "The Shining Ones," Clarke defines a furlong, which is, in fact, an eighth of a mile, somewhat large for giant squid as we know them, but he writes that Melville "was a man who met sperm whales every day, groping for a unit of length to describe something a lot bigger . . . so he automatically jumped from fathoms to furlongs."

"Big Game Hunt" is only a six-page story, concerning a biologist who has devised a method for electrically affecting the behavior of various invertebrates and goes to sea to try his device on our old friend *Architeuthis*—here called *Bathyteuthis* by Clarke. It works, the squid is summoned to the surface by electrical impulses, and they even film the "monstrous beast that no human being had ever before seen under such ideal conditions." Unfortunately, the squid is more powerful than the device, and a blown fuse results. As soon as the squid realizes that it is "its own master," it reacts violently, and although the end of the story is largely left to the reader's imagination, we have to assume that the film and the professor were victims of a very angry, very large squid.

In *The Deep Range,* for some reason, Clarke refers to the animal as *Bathyteuthis maximus,* although he certainly knew its proper scientific name. Set in the future, when men are farming whales for profit, this novel concerns one of the whale-rangers, who discovers that sperm whale casualties are unusually high in a particular sector, and one whale, found badly mauled and dead at the surface, is covered with sucker marks 6 inches in diameter. The scientists and rangers realize that the only creature capable of inflicting such damage on a sperm whale is a giant squid that they suggest "may be a hundred and fifty feet long." But instead of merely killing the monster, the men decide to capture it. Using a light-studded submarine as a lure, they descend to a depth of six hundred fathoms:

> A forest was walking across the sea bed—a forest of writhing, serpentine trunks. The great squid froze for a moment as if impaled by the searchlights; probably it could see them, though they were invisible to human eyes. Then it

gathered up its tentacles with incredible swiftness, folding itself into a compact, streamlined mass—and shot straight upward toward the sub under the full power of its own jet propulsion.

They manage to implant a sonar beacon in the mantle of the giant squid—nicknamed Percy—so they can return to capture him. They have been offered fifty thousand dollars by Marineland for the delivery of a healthy giant squid ("a giant squid would be the biggest attraction Marineland ever had"), and they plan to use the money to further their research. (At this point in the story, Clarke again quotes the entire giant squid passage from *Moby-Dick*.) Percy measures "a hundred and thirty feet from his flukes to the tips of his feelers," and they manage to anesthetize him with narcotic bombs, which enable them to bring him to the surface, draped like a limp dishrag over one of the subs. They install him in a specially built pen, and the first giant squid ever brought to the surface is now in captivity:

> First it swam slowly from end to end of the rectangular concrete box, exploring the sides with its tentacles. Then the two immense palps started to climb into the air, waving towards the breathless watchers gathered around the edge of the dock. They touched the electrified netting—and flicked away with a speed that almost eluded the eye. Twice again Percy repeated the experiment before he had convinced himself that there was no way out in this direction, all the while staring up at the puny spectators with a gaze that seemed to betoken an intelligence every wit as great as theirs.

Would that Clarke had followed up on Percy's captivity. Did they get him to Marineland? What did they feed him? Did he live happily ever after? Unfortunately, *The Deep Range* abandons Percy in his pen, and goes on to the conclusion of the story, in which the leader of the Buddhist world wants to eliminate whale farming altogether, and the man who actually captured Percy has to contend with bigger issues than a squid in a concrete box. But in 1992, at the conclusion of his *Omni* article, Clarke hopes that Benchley's *Beast* "does not trigger another ocean pogrom, aimed at the giant squid." (Clarke holds Benchley responsible for the massive destruction of white sharks following the publication of *Jaws*.)

Clarke's giant squids are a far cry from the malevolent monsters envisioned by Jules Verne, H. G. Wells, and Peter Benchley. The animal Clarke calls *Bathyteuthis* in one story is larger than anyone else's giant squid, but instead of attacking and eating people, it only wants to be left alone in the depths to light up and avoid sperm whales. Even though the only specimens we have seen outside of fantasy have been dead or dying, and have given no clue as to their normal behavior, Clarke persists in his ardent admiration. He writes, "This century has seen a complete transformation in our attitude towards other animals, including many once considered implacably hostile. . . . The giant squid is almost certainly a highly intelligent animal; given the opportunity, it might be as playful as its cousin—that charming mollusk, the octopus." Then again, it might not.

If you write an entire novel about a sperm whale, somewhere you will have to insert an episode about an encounter with a giant squid—after all, the protagonist has to *eat*. (Melville certainly wrote a novel that has a lot to do with a certain whale, but I think it is fair to say that *Moby-Dick* is not actually *about* a sperm whale.) In Hank Searls's 1982 *Sounding*, a novel in which sperm whales do a lot of thinking about other whales and why men are killing them, there comes a time when one of the whales gets hungry:

> Almost two thousand feet below her, in the abyss off the continental shelf, she echoed on eight giant squid, the largest she had scanned for years. . . . They were near her depth limit. Though she was forty feet long, the smallest squid's two largest tentacles almost equaled the length of her body. Its eight sessile members, the shorter arms with which the squid grasped its food, were a good ten feet in length. The seven other squid were out of the question. Squid so large might easily cap her blowhole and drown her below. . . . Any giant squid at its own depth had advantages in speed and eyesight over the sperm whale. It carried a gleam of luminescence that, though primarily for prey attraction, enabled it sometimes to glimpse by reflected light. The squid itself was a murderous carnivore. Its ten arms were dappled with tightly gripping suckers, each with a sawtooth edge that could cut through blubber like a flenser's blade. The two longer of these arms hugged the victim to the shorter ones, which were studded with suckers from tip to base and as strong as tempered steel. A thousand-pound squid in normal feeding could latch its longer members to an arctic shark, draw

it struggling to the shorter arms, and squeeze it immobile as it tore it to bits with its parrot beak and devoured it under cover of its mantle.

Searls, who also wrote *Jaws 2*, was as fascinated by "murderous carnivores" as Peter Benchley, and liked describing their modus operandi, even if he had to make it up. The female sperm whale "blanged" at the squid with a burst of sound, and "stunned him, but not enough. He slithered half-free, and she felt his long suckered tentacles groping for her eyes and his beak probing for her blowhole. . . . She heaved for the surface, burdened by eight hundred pounds of grasping, slicing squid."

It is not altogether surprising to find a giant squid in E. Annie Proulx's *Shipping News,* a novel based in Newfoundland. It appears only as a sidebar to a story about one of the men who is—like almost everyone else in this tale of modern Newfoundland—down on his luck:

"Seems like he's marked. He's the one had a fright, eight, nine years ago. Turned his hair snow white in a month. He was out fishing, see, with his brother near the Cauldron, and sees this limp old thing lying in the water. He thought it was a ghost net, you know, broke loose and come up to the surface. So to it they goes, he gives a poke with his hook, and dear Lord in the morning, this great big tentacle comes out of the water—" Dennis held his arm above his head, hand curved and menacing, "and seizes him. Seizes him around the arm. He says you never felt such strength. Well, lucky for him he wasn't alone. His brother grabs up the knife he was using to cut cod and commences sawing at the gripping tentacle, all muscle and suckers clamped tight enough to leave terrible marks. But he cut through it and got the motor started, his heart half out of his mouth expecting to feel the other tentacles coming down on his shoulder. They was out of there. The university paid them big money for that cut-off tentacle."

This brings to mind the story of another Newfoundlander, Theophilus Piccot. True or not, the story of Piccot and the giant squid of Portugal Cove (see pages 82–84) has become a part of Newfoundland's history, and once again the inclusion of *Architeuthis* rewards the author; Proulx won the Pulitzer Prize and the National Book Award for this 1993 novel.

Don Reed, a former diver at Marine World Africa USA in California, turned to writing books for "young adults" after his oceanarium career ended in 1987, and continues to write about the sea. In 1995 he published *The Kraken*, a book that has young Tom Piccot as its protagonist, and you-know-who as the threatening presence. In researching the book, Reed actually went to Newfoundland and met with various descendants of Tom Piccot, and also with Margueritte Aldrich, the widow of Frederick Aldrich, who "talked squid" with him. Except for the event with the giant squid, little is known about Tom Piccot, so Reed fleshes out the story—as is his right as a novelist—with a dramatic early encounter with the giant squid of Portugal Cove, where the squid rises to the surface and steals a fish that Tom is hauling in:

> Something *else* was rising underneath the halibut. Something so huge a bulge of water came before it, swelling the surface as if a great whale was rising. Just a school of herring, Tom told himself, whole bunch of 'em, that's why it looks so big. He waited for the school to change direction. But it was not many fish; it was only one creature. And it did not turn away. . . . Looking bigger and bigger as it neared, the creature below him changed colors: now white, now purplish pink, finally a hue unrecognizable in the dark, but which Tom knew was deepest red, the color of blood and rage. . . . It was . . . a squid, a giant squid, five times the length of his boat. The eyes were huge, black and dead white, and flickering a hideous green.

Many other Newfoundland (and some non-Newfoundland) events take place, including cod fishing, the giant squid attack on the schooner *Pearl* (which, if it took place at all, occurred in the Indian Ocean; see pages 198–201), nasty storms, sealing, trapping, and hunting, but, of course, the denouement of the novel occurs as young Tom Piccot hacks off the tentacle of the great squid that attacks the boat after they row up to it. He even brings the tentacle to Moses Harvey (the man who actually received the tentacle in 1873), a preacher in St. John's, who becomes slightly hysterical about his acquisition: "Nineteen feet six inches long, more than three times the height of a man! Ha-haaa! At last! At last! My boy, you don't know what this means! With this"—he clutched the tentacle tip, thrust it at Tom—"I hold in my hands the key to a mystery! For the first time the world has proof—positive, undeniable proof! The devilfish, the kraken, the

The cover of Peter Benchley's 1991 novel. Each of the suckers on the arm is drawn with a single claw in the middle, nothing like the real thing.

giant squid exists!" It certainly exists in Reed's modest but well-researched novel, and in a larger and considerably more flamboyant form in Peter Benchley's *Beast.*

In this 1991 novel, Benchley imagined his "beast" as a powerful, vengeful hunter and, despite all evidence to the contrary, made it 100 feet long. In fiction, a giant squid can be 100 feet long; it can weigh a dozen tons; it can blink its green lights on and off; it can have hooks on its tentacles; it can even feel anxiety. Real ones are not so cooperative in revealing anything but their gross morphology—and then so inadequately that we still don't know how many species there are, where they live (we only know where they die), what they eat, how they breed, how big they get, or whether they are aggressive hunters or passive eaters of carrion.

Like all of Benchley's "man-eating monster" novels, *Beast* became a bestseller.* *Jaws* was a blockbuster when it was published in 1974, but it did not enter the stratosphere until it was made into a film, directed by Steven Spielberg. Before the 1977 release of *Star Wars, Jaws* was the highest-grossing movie in Hollywood history. Benchley had intended to call his giant squid novel *The Last Monster,* but his publisher thought that *Beast* would be a better name, and *Beast* it became. (In the novel, however, he has an expert on giant squid write a book called *The Last Dragon.*)

---

*In 1994, he wrote *White Shark,* which is not about a shark at all, but rather about an experiment conducted by the Nazis before the end of the war, in which a water-breathing human was designed as a weapon of war, equipped with gills and stainless-steel teeth. The German code name for this monster was *Weisshai,* which means "white shark."

Just as in *Jaws,* the book opens with a description of the eponymous monster, lurking in the darkness of the ocean:

> It hovered in the ink dark water, waiting.
>
> It was not a fish, had no air bladder to give it buoyancy, but because of the special chemistry of its flesh, it did not sink into the abyss.
>
> It was not a mammal, did not breathe air, so it felt no impulse to move to the surface.
>
> It hovered.
>
> It was not asleep, for it did not know sleep, sleep was not among its natural rhythms. It rested, nourishing itself with oxygen absorbed from the water pumped through the caverns of its bullet-shaped body.

And again, as in *Jaws,* the first thing the monster does is gobble up some unsuspecting humans. The similarities between the shark thriller and the squid thriller might lead one to suspect a certain dependence on formula: big monster begins eating people in beach town, panic ensues, hero (often marine biologist accompanied by valiant fisherman) offers to subdue monster; town leaders do not want to close beaches, more people get eaten, intrepid hero triumphs and dispatches monster in a burst of pyrotechnics. There is even a delicate self-referential episode in *Beast* in which Whip Darling (the valiant fisherman) says, "Whenever I hear talk about monsters, I think about *Jaws.* People forget *Jaws* was fiction, which is another word for . . . well, you know, B.S. . . . My rule is, when someone tells me about a critter as big as a tractor-trailer, I right away cut a third or a half off what he says. . . . But with this beast, seems to me when you hear stories about him, the smart thing to do is not cut anything off. The smart thing to do is double 'em."

It is not surprising that Benchley chose *Architeuthis* as the protagonist of his novel. Given the (largely erroneous) hyperbole that has circulated about the giant squid, *Architeuthis* certainly qualifies as a scary monster. There is something inherently odious about a 60-foot-long creature with a fierce beak, gigantic, unblinking eyes, and flailing arms equipped with claws and suckers.

Many of the elements in *Beast* came from Benchley's imagination, but some—like the story he tells of "something" dragging fishermen's traps around the bottom at great depths—have a basis, if not in fact, then in legend. As reported by J. Richard Greenwell in the International Society of Cryptozoology *Newsletter,* a Bermuda fisherman named John P. "Sean" Ingham was having a lot of trouble with his traps. Fishing for deep-water crabs and shrimps, he regularly lowered to between one thousand and two thousand fathoms; these specially reinforced traps had been coming up bent and damaged, or, in some cases, not coming up at all.

Then, on September 3, 1984, Ingham was winching up a trap that had been on the bottom at five hundred fathoms. About halfway up, the line broke. It would take a weight of over six hundred pounds to break the line, which was polyethylene rope. Dr. Bennie Rohr, a biologist with the National Marine Fisheries Service Southeast Laboratory at Pascagoula, Mississippi, suggested that it might be the work of a giant octopus, since a full cage of tasty shrimps or crabs would be the perfect bait for it. (Giant squid presumably feed on faster-moving prey, such as fishes and other, smaller, squids.) On another occasion, when a smaller trap was being hauled up from 480 fathoms, the trap seemed rooted to the bottom as if something very heavy was holding it, and the line, with a forty-five-hundred-pound breaking strain, was beginning to give. As if to prove that the trap was not snagged on the bottom, whatever was holding it began to pull the fifty-foot boat. When Ingham put his hand on the line, he felt "thumps like something was walking."

The eponymous "beast" in Benchley's novel turns out to be a 100-foot-long, anthropophagous giant squid, but the attacks on the fish traps of Whip Darling are straight out of Sean Ingham's logbook. In the novel, Darling is pulling his traps in deep water when the mate, Mike, tells him "Something's not right." At Mike's suggestion, Darling puts a hand on the rope, which is "trembling erratically. There was a thud to it, like an engine misfiring." As the line is hauled up, the stainless-steel cable that holds the cage appears at the surface, but the trap is gone. "Bit," says Darling. "Bit clean through." Four more traps are pulled, and four more cables are bit clean through. The last of the cables comes up whole, but the trap has been wrapped around its weights so hard "it was as if everything had been melted together in a furnace."

Mike stares at it for a long moment, then says, "Jesus, Whip. What kind of sumbitch do that?"

"No man for sure," says Darling. "No animal, neither. At least no animal I've ever seen."

In *Sharks Are Caught at Night*, published in 1958 and purporting to be a true account of his adventures in the Caribbean, François Poli tells a remarkably similar story, which, he says, "was front page news in every Cuban paper." A fisherman named Sanchez found that his buoys were being pulled down "slowly and without jerks," and when he could not raise his lines he realized "that his hook had not caught on a rock but in a living and incredibly heavy mass." He knew it was not an octopus, because "an octopus wouldn't have pulled like that." For three days they tried to capture the monster, without success.

When the fishermen told their story, it was picked up by the press, and "two days later fifty thousand Cubans learned as they ate their breakfast that a huge monster was roaming around the island at a depth of 100–150 fathoms." A fisherman named Torial offered an explanation. He said that "off certain desolate coasts in Mexico monsters appeared which had never been accurately described, as no one had ever come within a mile of them. They had a huge cylindrical body striped with yellow, and tentacles something like those of an octopus. Whenever one of them was reported off-shore the fishermen refused to put to sea for days." In the end, nobody ever figured out what the creature was, but when Ralph Thompson illustrated the English edition of Poli's book (it was originally published in French), he drew the mystery beast as a giant squid.

"Its eight sinuous arms floated on the current," wrote Benchley in *Beast*, "its two long tentacles were coiled tight against its body. When it was threatened or in the frenzy of a kill, the tentacles would spring forward, like tooth-studded whips. . . . It existed to survive. And to kill. For, peculiarly—if not uniquely—in the world of living things, it often killed without need, as if Nature, in a fit of perverse malevolence, had programmed it to that end."

Is any of this legitimate? Except that a squid has eight arms and two longer tentacles, not a single word is accurate. Even if we knew about the giant squid's eating habits—which we don't—it is absurd to say that it "often killed without need." Its "tooth-studded whips" are also imaginary (but necessary to the story), since *Architeuthis* does not have teeth on its tentacles—or anywhere else, for that matter, unless you count the microscopic denticles on its radular tongue.

Despite the intentions of the authors (or perhaps *because* of them), the line between fact and fiction is often obliterated in these "monster" novels—think of how people reacted to the stories of the homicidally maniacal white shark in *Jaws*—and sensationalism often supersedes reality. One might argue that a writer of horror fiction is not bound by the niceties of biological accuracy. The same argument could be applied to Jules Verne, who wrote that his 25-foot-long giant squid weighed between 40,000 and 50,000 pounds, and slid its arms down the hatchways of the *Nautilus*, plucking hapless sailors out of the submarine with the intention of devouring them. Nevertheless, it is grating—at least to those who understand how little is actually known about *Architeuthis*—to hear the "marine biologist" in *Beast* spouting such nonsense.

After two scuba divers are gobbled up, the town fathers call in Dr. Herbert Talley, the world's foremost authority on squids and the author of *The Last Dragon*, the definitive work on the subject. When Dr. Talley arrives in Bermuda, he tells Whip that he is "a doctor of malacology," which, he explains, means "a doctor of squid." Dr. Squid delivers a lecture about the habits of his favorite subject in which he asserts that *Architeuthis* is "what we call an adventitious feeder. He feeds by accident, he eats whatever's there. His normal diet—I've looked in their stomachs—is sharks, rays, big fish. But he'll eat anything." (During the same lecture, Talley tells Whip that the squid are around because their normal food items have been fished out of Bermuda waters, and besides, their only predators, the sperm whales, "could already be practically extinct.")

As the story builds to its final confrontation, the men are at sea, having decided to lure the squid into range with a sex decoy—a hook-studded, squid-shaped lure that releases "a chemical that perfectly replicates the breeding attractant of *Architeuthis*"—so they can shoot it or blow it up with Semtex, a plastic explosive. The squid, drawn inexorably by the "most basic of all impulses," grabs the lure, dashes off, and then, when it realizes that it has been fooled by a poor imitation, it gets *really* mad. It returns to the ship, and because the would-be squid hunters had the foresight to deploy an underwater video camera, they see their cephalopod adversary for the first time. Talley looks at its eye and whispers, "The thing must be ninety feet, perhaps more. . . . This could be a hundred-foot animal."

Adapted from Peter Benchley's novel about a giant squid, the made-for-television movie The Beast arrives in New York.

The squid attacks the boat, furious that these puny humans dared to trifle with its emotions:

Its chemistry was agitated, and its colors changed many times as its senses struggled to decipher conflicting messages. First there had been the irresistible impulse to breed; then perplexity when it had tried to mate and been unable to; then confusion when the alien thing had continued to emit breeding spoor; then anxiety as it had tried to shed the thing and found that it could not.

Even though squids and octopuses are considered the intellectuals of the cephalopod family, this is a rather exceptional range of emotions for an invertebrate whose close relatives include clams, mussels, and snails. But this is one furious squid, and as it attacks the ship, it loses one of its arms to the whirling propeller. Does the loss of an arm discourage it? Hardly. It reaches up on the deck with one of the club ends of its whip:

Something was coming over the bulwark. For a moment it seemed to ooze like a giant purple slug. Then the front end of it curled back like a lip, and it began to rise and fan out until it was four feet across and eight feet high, and it blacked out the rays of the sun. It was covered with quivering circles, like hungry mouths, and in each one Darling could see a shining amber blade.

With its seven arms and two whips, the squid is grabbing everything on the deck of the ship, and it has poor Dr. Talley as an hors d'oeuvre. When Whip Darling tries to defend himself with a boat hook, the squid delicately removes it from his hands and drops it into the sea. Then Darling tries a chain saw, with the same results. All seems lost, and with everybody snared in a tangle of writhing arms, along comes the mother sperm whale (the *deus ex Physeter*?) whose calf had been killed by the hungry squid several chapters earlier. She leaps out of the water like Flipper, and bites the squid's head off. The End?

Well, not quite. *Beast* (the movie) is roughly the same as *Beast* (the novel), but with the obligatory emendations for Hollywood purposes. (The movie was actually shot in Australia.) In the novel, Whip Darling is a Bermuda fisherman; in the film, he becomes "Whip Dalton," and the venue has been moved to Washington State (where, incidentally, there has never been a report of a giant squid). Instead of a single, hungry beast, there are now two, mother and child. Same old business about the town fathers not wanting the bad publicity, and more old business about nobody believing that such a creature could exist. ("Nothing could get to be that big," etc.)

As in the novel, the squid picks off unwary sailors, and leaves behind a claw as evidence of its existence. But where the claw in the novel is described as 2 inches long, the one picked up by Whip Dalton is about 5 inches long, resembling nothing more than one of the eviscerating talons of the velociraptors in *Jurassic Park*. In the movie, we see the squid, moving menacingly through the water as it prepares for its next meal. Most of the depictions of the squid appear to be accurate—as far as we know, since no one has ever actually seen one—but for some reason, they have the squid lying on the bottom as the two scuba divers approach their demise. (Giant squid may indeed lie on the bottom, but if they do, they are the only squid with that inclination.)

The squid hunters realize that the 37-footer is "only three months old" (some growth rate!) and that the mother is around and really angry. ("*Architeuthis* is known to be a vengeful beast," says Talley, "it was interested in only one thing—

it wanted to kill . . . and here, my God, its child has been murdered. How could it not want vengeance?") They set out with the sex decoy, but Manning, the aquarium owner, who recognizes the dollar value of a living giant squid, plans to anesthetize it instead of killing it, using phenobarbital instead of cyanide, and bring it back alive to Texas. The squid takes the lure, and as they reel it in, Manning manages to shoot a couple of his tranquilizing darts into it. Since the others thought he was going to kill it, they celebrate its demise—somewhat prematurely, as it turns out. Everybody not in the aquarium business worries about what will happen when the snoozing cephalopod wakes up, and they don't have long to wait. The phenobarb wears off, and screaming lustily, the squid breaks the "unbreakable" cable that held it, and prepares for the denouement.

Screeching like a demented parrot, Momma Beast attacks Whip's vessel, and forsaking the ecologically correct resolution of the novel, the moviemakers provide no sperm whale to save them. With her arms flailing and grabbing, the squid knocks Manning into the water and drags poor Dr. Talley toward her snapping beak. She grasps Whip's leg while Lieutenant Marcus dangles from a ladder lowered by the Coast Guard helicopter. "Shoot the flare! Shoot the flare!" Whip shouts, and Lieutenant Marcus does, igniting the gasoline that has spilled on the deck in the melee. The boat blows up, the squid becomes the largest grilled cala-

Pierre Denys de Montfort's 1802 illustration of some sort of a giant cephalopod attacking a ship.

mari in history, and Whip, who has managed to climb the ladder too, survives to fish again—if his boat insurance covers giant squid attacks.

In dealing with a creature as mysterious and problematical as *Architeuthis,* one frequently encounters a unique category of written material: fiction that the author wants you to believe is fact. This is completely different from well-written fiction, where you know it is not true but you suspend your disbelief long enough to enjoy the story. In this category, we are presented with accounts by supposedly reputable people and asked to believe them, often with no corroborative evidence save for the storyteller's assurance that the story is true. This technique is frequently used in the practice of cryptozoology, as in, for instance, the assorted "sightings" of the Loch Ness Monster. (The familiar "surgeon's photograph," supposedly taken by a person whose credentials were unassailable, turns out to have been a fake.) The giant squid is so elusive, and its natural history so mysterious, that evidently some people simply cannot resist the temptation to make up stories—and then try to convince others that they are true.

The French naturalist Pierre Denys de Montfort published the *Histoire naturelle générale et particulière des mollusques* in 1802, which, in one form or another, was probably responsible for much of the early lore about the *poulpe.* Off the coast of Angola, goes the story, a sailing ship was seized by a monster with arms that reached to the top of the masts. A picture of this scene appears in de Montfort's book. The terrified sailors vowed to St. Thomas that they would make a pilgrimage if he would save them, and with axes and cutlasses—and the support of the good St. Thomas—they broke free. Thereafter (according to de Montfort), a votive picture of a ship in the embrace of a monster was placed in St. Thomas's chapel at St.-Malo. We do not know if Jules Verne ever saw this picture—if it really existed—but there is hardly any question that Verne read de Montfort's study. The original illustrator of *Vingt mille lieues sous les mers (Twenty Thousands Leagues Under the Sea),* identified only as "de Neuville," also must have seen the reproduction in de Montfort, for he used a modified version in Verne's book. The one notable difference was that where de Montfort depicts an eight-armed animal that could be a squid, de Neuville's version has obvious octopus overtones. (It is known that an octopus was displayed at the aquarium at Boulogne as early as 1867, so the author, the artist, or both could have seen it.)

In the chapter "Gigantic Cuttle-Fishes" in *The Octopus* (published in 1875), Henry Lee calls the de Montfort illustration "fitter to decorate the outside of a showman's caravan at a fair than seriously to illustrate a work of natural history." (A. S. Packard, writing in the *American Naturalist* in 1873, refers to "the well-known hoax of Denys Montfort.") Nevertheless, the illustration gained credence, thanks to Verne.

De Montfort's work certainly had an impact on people's perceptions of the creature's size. In his 1958 study of marine monsters, subtitled *Le kraken et le poulpe colossal*, Bernard Heuvelmans presents probably the most detailed biography of Denys de Montfort in existence. (The chapter is called "Pierre Denys de Montfort, malacologue maudit"—"accursed malacologist.") In researching his *Histoire naturelle des mollusques*, de Montfort evidently interviewed a Yankee whaling captain (identified by Heuvelmans as "Benjohnson") who told him about a sperm whale they captured that had something very curious sticking out of its mouth. Heuvelmans quotes de Montfort quoting Benjohnson:

> They could hardly believe their eyes when they saw that this fleshy mass, truncated at both ends, the thickest of which was as big as a mast, was nothing but the arm of an enormous octopus, whose sunken suckers were broader than a hat; the lower end appeared freshly severed; the upper end must have been cut in some struggle which had preceded its capture by some time, for it was scarred and extended by a kind of extension the size and length of a man's arm. This limb of the enormous octopus, accurately measured with a fishing line, turned out to be seven fathoms, or 42 feet long and the suckers were arranged in two ranks as in the common octopus.

From this description, Denys de Montfort decided that another 10 feet was missing from the upper end of the arm and another 20 feet from the lower, which would have made it 72 feet long. Another story told to de Montfort involved a whaler named Reynolds whose men found a 45-foot-long arm floating on the surface after they harpooned a sperm whale; they brought a piece of it on board, hung it for a while, and then ate it.

According to Henry Lee, it was de Montfort's purpose to "cajole the public," and he is reported as saying, "If my entangled ship is accepted, I will make my 'colossal poulpe' overthrow a whole fleet." Both events seem to have occurred, for soon de Montfort was putting forth the story that six French men-of-war and the four British ships that captured them were attacked and sunk by a colossal cuttlefish. Heuvelmans concedes that de Montfort told these stories, but claims that he was only joking. Ostracized and ridiculed, de Montfort retired to the country, where he wrote a pamphlet about beekeeping. Lee concludes his discussion of de Montfort by saying, "I have been told, but cannot vouch for the truth of the report, that de Montfort's propensity to write that which was not true, culminated in his committing forgery, and that he died in prison." According to Heuvelmans, de Montfort sank deeper and deeper into poverty, until he starved to death in 1820 or 1821.

Jules Verne probably knew of the available published material on giant squid before he wrote his own book. In fact, he incorporates much of it into *Twenty Thousand Leagues Under the Sea*. He refers by name to Olaus Magnus and Bishop Pontoppidan, and then delivers a brief discourse on events involving the ship *Alecton,* where a giant squid was encountered at sea, and "Commander Bouguer"

> had the creature harpooned and shot, but without much success, for the bullets and the harpoons went through its soft flesh as if it were loose jelly. After several unsuccessful attempts, the crew managed to pass a slipknot around the creature's body. The loop slid back to the caudal fins and there it stopped. They then tried hauling the monster on board, but it weighed so much that the tail merely broke off and the squid itself, separated from this part of its body, disappeared beneath the surface.

There are few authenticated records of a giant squid attacking a ship, let alone a man, and those that do exist are open to question. In one account, printed in the London *Times* for July 4, 1874, the master of the steamer *Strathowen,* bound from Mauritius to Rangoon, sighted a small schooner (later identified as the *Pearl*) and, next to it, "a long, low swelling lying on the sea, which, from its colour and shape, I took to be a bank of seaweed." The "seaweed" reached up and dragged the schooner under, and several members of the crew escaped and were picked up by the *Strathowen.* Here is the article, in its entirety, as it was published in the London *Times:*

The following strange story has been communicated to the Indian papers:

"We had left Colombo on the steamer Strathowen, had rounded Galle, and were well in the bay, with our course laid for Madras, steaming over a calm and tranquil sea. About an hour before sunset on the 10th of May we saw on our starboard beam and about two miles off a small schooner lying becalmed. There was nothing in her appearance or position to excite remark, but as we came up with her I lazily examined her with my binocular, and then noticed between us, but nearer her, a long, low, swelling lying on the sea, which, from its colour and shape, I took to be a bank of seaweed. As I watched, the mass, hitherto at rest on the quiet sea, was set in motion. It struck the schooner, which visibly reeled, and then righted. Immediately afterwards, the masts swayed sideways, and with my glass I could clearly discern the enormous mass and the hull of the schooner coalescing—I can think of no other term. Judging from their exclamations, the other gazers must have witnessed the same appearance. Almost immediately after the collision and coalescence the schooner's masts swayed towards us, lower and lower; the vessel was on her beam-ends, lay there a few seconds, and disappeared, the masts righting as she sank, and the main exhibiting a reversed ensign struggling towards its peak. A cry of horror rose from the lookers-on, and, as if by instinct, our ship's head was at once turned towards the scene, which was now marked by the forms of those battling for life—the sole survivors of the pretty little schooner which only 20 minutes before floated bravely on the smooth sea. As soon as the poor fellows were able to tell their story they astounded us with the assertion that their vessel had been submerged by a giant cuttlefish or calamary, the animal which, in a smaller form, attracts as much attention in the Brighton Aquarium as the octopus. Each narrator had his version of the story, but in the main all the narratives tallied so remarkably as to leave no doubt of the fact. As soon as he was at leisure, I prevailed upon the skipper to give me his written account of the disaster, and I have now much pleasure in sending you a copy of his narrative:"

"I was lately the skipper of the *Pearl* schooner, 150 tons, as tight a little craft as ever sailed the seas, with a crew of six men. We were bound from Mauritius for Rangoon in ballast to return with paddy, and had put in at Galle for water. Three days out, we fell becalmed in the bay (lat. 8°50'N, long. 85°05'E). On the 10th of May, about 5 P.M.—eight bells I know had gone—we sighted a two-masted screw on our port quarter, about five or six miles off, very soon after, as we lay motionless, a great mass rose slowly out of the sea about half-a-mile off on our larboard side, and remained spread out, as it were, and stationary; it looked like the back of a huge whale, but it sloped less, and was of a brownish colour; even at that distance it seemed much longer than our craft, and it seemed to be basking in the sun. 'What's that?' I sung out to the mate. 'Blest if I knows; barring its size, colour, and shape, it might be a whale' replied Tom Scott; 'and it ain't the serpent,' said one of the crew, 'for he's too round for that 'ere critter.' I went into the cabin for my rifle, and as I was preparing to fire, Bill Darling, a Newfoundlander, came on deck, and, looking at the monster, exclaimed, putting up his hand, 'Have a care, master; that 'ere is a squid, and will capsize us if you hurt him.' Smiling at the idea, I let fly and hit him, and with that he shook; there was a great ripple all round him, and he began to move. 'Out with all your axes and knives,' shouted Bill, 'and cut at any part of him that comes aboard; look alive, and Lord help us!' Not aware of the danger, and never having seen or heard of such a monster, I gave no orders, and it was no use touching the helm or ropes to get out of the way. By this time three of the crew, Bill included, had found axes, and one a rusty cutlass, and all were looking over the ship's side at the advancing monster. We could now see a huge oblong mass moving by jerks just under the surface of the water, and an enormous train following; the oblong body was at least half the size of our vessel in length and just as thick; the wake or train might have been 100 feet long. In the time I have taken to write this the brute struck us, and the ship quivered under the thud; in another moment, monstrous arms like trees seized the vessel and she heeled over; in another second the monster was aboard, squeezed in between the two masts, Bill screaming 'Slash for your lives,' but all our slashing was of no avail, for the brute, holding on by his arms, slipped his vast body overboard, and pulled the vessel down with him on her beam-ends; we were thrown into the water at once, and just as I went over I caught sight of one of the crew, either Bill or Tom Fielding, squashed up between the masts and one of these awful arms, for a few seconds our ship lay on her beam-ends, then filled and went down; another of the crew must have been sucked down, for you only picked up five; the rest you know. I can't tell who ran up the ensign.

"James Floyd, late master, schooner *Pearl*."

The incident—if it occurred—took place soon after the publication of *Twenty Thousand Leagues*. Bernard Heuvelmans, who wanted very much to believe in sea monsters (and wrote a book called *In the Wake of the Sea-Serpents*), questions the veracity of the account, writing, "This tale has never been confirmed and it may well have been an opportune hoax, for the *Strathowen* is not to be found in *Lloyd's Register* for that year." Frank Lane also investigated the story, but could find no confirmation in Britain—"from Lloyd's, the National Maritime Museum, the General Register of Shipping and Seamen, shipping lines and other likely sources." Nevertheless, he chose to accept it, and in his 1963 work wrote, "The most *reasonable* explanation seems to be that the account was a report of an actual incident, including the presence on the *Pearl* of a man from the one place [Newfoundland] where, at that time, giant squids and their behavior were reasonably well-known."

When a story appears in a reputable journal, it is more likely to be believed than if it appears in the popular press. When Arne Grønningsaeter's fifteen-thousand-ton freighter *Brunswick* was sailing between Hawaii and Samoa and was "attacked" by a giant squid, it was reported in the Norwegian journal *Naturen* (Nature) in 1946. The event occurred in the Pacific between 1930 and 1933 and was recounted by Grønningsaeter, the master of the ship. Although short on details (he never even estimated the size of the squid), Grønningsaeter's account described the squid swimming alongside the ship at a speed of twenty to twenty-five knots, and then turning toward the vessel, "hitting the hull approximately 150 feet from the stern at a depth of 12–15 feet." Since it could not get a grip on the hull, it "skidded along until it ended up in the propeller, where it was ground to pieces." If this occurred, it is the only such story in all the literature, a circumstance that raises the question of its authenticity. If giant squid attack ships, why did it happen only once?

When there is no physical evidence to corroborate a giant squid report, we often have to rely on anecdotal information, which is often exaggerated and occasionally altogether unbelievable. Nevertheless, some of these stories make it into the popular literature, where they contribute to the already opprobrious reputation of *Architeuthis*.

Paul LeBlond and John Sibert, both respected zoologists in British Columbia (and both dedicated cryptozoologists), assembled a sixty-four-page report entitled "Observations of Large Unidentified Marine Animals in British Columbia and Adjacent Waters." Mostly devoted to British Columbia's abundant sea serpents,* this unpublished report also contained a couple of references to giant cephalopods, such as the story communicated to them (complete with drawing) from a Mr. Charles Dudoward of the "great squid which washed ashore one winter morning of 1922 in front of Mrs. Robertson D. Rudge's Port Simpson Hotel." The drawing (a photocopy of the pastel sketch that was submitted to LeBlond and Sibert) showed a large cephalopod stretched out on the front lawn of the hotel. It was described as "having four long arms on each side, 50 ft. long . . . but the one in the middle is about 100 feet long and might be longer when stretched." LeBlond and Sibert, in their comments on this creature, wrote, "This report is very reminiscent of the numerous descriptions of giant squids found on Newfoundland's beaches at the end of the last century. . . . The description fits *Architeuthis* very well and one may safely conclude that this was indeed a specimen of the giant squid."

There is no reason to doubt Mr. Dudoward's report, even though it comes to us as a secondhand story from an unlikely region; as LeBlond and Sibert write, "To our knowledge, this is the only specimen ever found in this region of the Pacific." In *There Are Giants in the Sea,* however, Michael Bright repeats the story, and even adds another of LeBlond and Sibert's twice-told tales, this one supposedly occurring in 1892, again near Port Simpson, which is on the coast of British Columbia, just south of Ketchikan, Alaska. It seems that a group of Indians were towing a log boom with a flotilla of fifty canoes when they were slowed down by some unseen force. When they finally managed to beach the raft, they discovered "an enormous squid, larger than the raft itself, squashed underneath. One arm was reported to be more than 30m (100 ft) long and it ended in a large hook. The suckers were described to be 'as big as basin plates, to saucer-sized at the ends.'"

---

*The most famous of British Columbia's sea serpents is *Cadborosaurus* (named for Cadboro Bay, and nicknamed "Caddy"), about which LeBlond and E. L. Bousfield have written an entire book. Published in 1995, *Cadborosaurus* details the sightings of Caddy, and reproduces many drawings by eyewitnesses. There are even photographs of what is said to be the carcass of a sea serpent taken from the stomach of a sperm whale in 1937.

(The "middle arm" of the 1922 squid also had a hook on the end.) Although Bright tells us that these are only stories, their appearance in his book legitimizes the reports and lends credibility to some of the more outlandish aspects—such as the 100-foot-long arms with hooks on the ends and suckers as big as basin plates. (Bright is senior producer of the BBC Natural History Unit in Bristol, and if we can't believe the BBC, whom can we believe?)

Along with the acknowledged roster of *Architeuthis* sightings and strandings, Bright has included many doubtful records, and while they may be legitimate, the lack of documentation means that we can either take his word that these things happened, or not. What we cannot do—because he has not told us where to look—is read the original reports and decide for ourselves if we wish to accept them.*

For example, what are we to make of his story about "a badly damaged carcass which locals claimed to be a giant squid, [that] was washed ashore at Port Shepstone on the Natal coast of South Africa in 1926. All its arms and tentacles were missing, but estimates, based on the size of the body alone, put its overall length, with outstretched tentacles, at about 30m (100ft)"?† Or: "Another specimen found at Flower's Cove on the Newfoundland coast in 1934 was positively identified and measured. It was 22m (72ft) long"? Or: "One found in the same area in 1882 was claimed to have a length of 26.9m (88ft)"? As far as I can tell, none of these specimens appears anywhere else in the literature, and if Mr. Bright has seen some sort of authentication of these records, he ought to tell us where to find it.

Bright includes a story of a giant squid that appeared alongside an Admiralty trawler lying off one of the Maldive Islands in the Indian Ocean. The witness was

---

There Are Giants in the Sea has hardly any references at all; the entire bibliography consists of nine books, four of which are pro sea serpent, four of which are about the identification of various sea creatures; the last is *The Guinness Book of Animal Facts and Feats,* by Gerald Wood.

†In Heuvelmans's 1958 book, there is a brief discussion of a gigantic carcass that washed ashore on October 25, 1924, "at Baven-on-Sea, near Margate." Margate is close to Port Shepstone on the Natal coast, so this may be the same event, slightly misdated. The reference for the Baven-on-Sea animal is a 1925 article in *Wide World Magazine* (London) that makes specific reference to an article "which recently appeared in the *Natal Mercury,* published at Durban." Heuvelmans sent me a copy of the article, but all my attempts to verify that such a story actually appeared in the Durban newspaper have failed.

J. D. Starkey, who often fished at night over the stern of the ship using a cluster of lightbulbs (the "bulb cluster" of the story) to attract the fish. Starkey sent his account to *Animals* magazine in response to an article that the magazine had run in September 1963 entitled "Is There a Sea Serpent?" One night, as he walked the deck on the midnight-to-4:00 A.M. "graveyard watch," he had an unusual visitor:

> The water appeared to become opaque as the bulk of something filled my view. As I gazed, fascinated, a circle of green light glowed in my area of illumination. This green unwinking orb I suddenly realized was an eye. The surface of the water undulated with some strange disturbance. Gradually, I realized that I was gazing at almost point-blank range at a huge squid.
>
> I say "huge"—the word should be "colossal," as so far all I could see was the body, and that alone filled my view as far as my sight could penetrate. I am not squeamish, but that cold, malevolent, unblinking eye seemed to be looking directly at me. I don't think I have ever seen anything so coldly hypnotic and intelligent before or since.
>
> I took my quartermaster's torch and, shining it into the water I walked forward. I climbed the ladder of the fo'c'sle and shone the torch downwards. There, in the pool of light, were its tentacles.
>
> As already explained, I would not exaggerate a natural phenomenon, but these were at least 24 inches thick. The suction discs could clearly be seen. The ends of the arms appeared to be twitching slightly, but this may have been a trick of the light.
>
> My heart was going like a sledgehammer. Remember, I was alone on the deck, everyone else turned in. I was not so much afraid as excited, as if this were an opportunity to see something rarely seen by man.
>
> I walked aft keeping the squid in view. This was not difficult as it was lying alongside the ship, quite still except for a pulsing movement. As I approached the stern where my bulb cluster was hanging, there was the body. Every detail was visible—the valve through which the creature appeared to breathe, and the parrotlike beak. Gradually, the truth dawned: I had walked the length of the ship, 175 feet plus. Here at the stern was the head or the body and at the bows the tentacles were clearly visible. . . . The giant lay, all its arms stretched along-

side, gazing up, first with one then with both eyes as it gently rolled. After 15 minutes it seemed to swell as its valve opened fully and without any visible effort it "zoomed," if I may use the expression, into the night.

I never told anyone aboard as I should have been scoffed at.*

This story is quite wonderful, if for no other reason than it describes a giant squid that is almost four times longer than anyone has ever documented. Indeed, there may actually be such monsters in the Indian Ocean, but so far, this is the only one that has ever been seen, and Bright concludes his discussion with a somewhat anticlimatic remark: "The largest authenticated giant squid found since 1900 was caught by the crew of a U.S. Coast Guard ship patrolling the Great Bahamas Bank near Tongue of the Ocean. . . . It measured 14.3m (47 feet)."

Of course, there are times when Bright *does* identify the source of his material: he quotes at some length from something called *Deep-Sea Bubbles* by Henry Hedger Bootes. Bootes writes that he shipped aboard a British whaler he refers to as *Anna Lombard* ("her real name does not matter"), in "188–," heading for the South Seas. Mr. Bootes will eventually describe an encounter with a giant cuttle-fish, but before that event, he treats us to such a display of biological and ceto-logical illiteracy that the reader must wonder if the writer ever went to sea at all, let alone on a whaler.† He describes a gigantic ray, 20 feet across, whose pectoral

---

*In a concluding note to Starkey's story, the editors of *Animals* wrote, "The giant squids are among the most remarkable, and at the same time, among the least known, of the creatures that live in the sea. From what little we know of them, they appear to inhabit the middle depths of the oceans (not the abyssal depths), but many deep sea creatures move towards the surface at night, and Mr. Starkey's observation suggests that these great mollusks may be among them. . . . A squid of the enormous size recorded by Mr. Starkey is not beyond the bounds of possibility, and we welcome the privilege of putting his remarkable adventure on record."

†My copy of *Deep-Sea Bubbles*, bought in a secondhand bookshop, was reviewed for the Explorers Club by the famous ornithologist and sometime whaleman Robert Cushman Murphy. (Murphy is the author of *Logbook for Grace* and *A Dead Whale or a Stove Boat*, both of which relate to his 1912 whaling voyage to South Georgia aboard the brig *Daisy*.) In the typed review, which is signed by Murphy and pasted into the back of the book, he says, "The author has read widely in many branches of science, but he gives no inkling that he has understood a word of it. Speaking as a naturalist, I can only say that his zoology is bunk unrelieved. . . . He uses a plethora of scientific names, but he usually applies them to the wrong animals. Others of his creatures could, in the nature of things, have neither Latin nor vernacular names, for he is the only man who has ever seen them."

fins are marked with brilliant scarlet and deep blue spots, which has "come to feast" on the sperm whale they have just harpooned; then he describes the "razor-back whale, also known as the rorqual *(Rorqualus australis),*" which uses its exten-uated dorsal fin to disembowel its victims. And when he discusses "a large herd of round-backed, long-snouted dolphins," he says, "I am sure the beauty of this fish is somewhat marred by the ugly long name that has been given it by some scientific 'busy-body.' Fancy calling a fish *Lagenorhynchus obliquidens.*" (That is actually the scientific name of the "fish" known as the Pacific white-sided dolphin.)

For some reason, Michael Bright chose to accept Bootes's fatuous account as legitimate. Because it includes a description of the cuttlefish, here it is reprinted with all its ludicrous misrepresentations:

One particular spot of the impenetrable depths assumed a silver-whitish appearance, which at times became quite luminous, and very gradually we made out the waving arms of a giant cuttlefish. It gathered speed as it rose and I saw the awful eyes, which seemed to fix their gaze on me, holding me speechless and perfectly spellbound. I was quite unable to take my eyes off it, and my mind went back to Madam S. and our talk of this monster, on the terrace of her house in Valparaiso. The waving tentacles and long snake-like arms, each with rows of suckers, claimed my attention. As they waved upwards I could see them opening and shutting in anticipation of a feast. The body would be about twenty feet across the middle, but great portions of heaving flesh seemed to encase the joints, or sockets of the arms and tentacles, giving it greater massiveness. For ugliness, nothing that the morbid imagination of man has ever invented can compare with this pulsating horror. I speculated in my mind as to which would fall a victim to the tentacles—our whaleboat, the floating carcass or the razorback; and although these thoughts flashed through my brain, I made no attempt to order my men to pull out of the danger zone. We were all more or less hypnotized and helpless.

Now I must say something of the razorback. We had made several attempts to beat it off, and each time it had dived under the carcass, tearing huge mouth-fuls of blubber from the stomach, and evading our harpoon as though accus-tomed to such sport; but as the cuttlefish came closer, it too came under the spell of the waving arms, or was it the fixed gaze of the eyes and the rows of teeth (on the radulla [sic]?) which protruded from the strange-looking jaws, resembling in appearance a collection of parrots' beaks, or crab claws. Presently, with one mighty spring, it seized the razorback, not anywhere near the dorsal formation, but round the small of the tail and neck. Then the water became impregnated with the sepia which this vile thing ejects, and secure in the entwining embrace of the octopus the rorqual was carried to the depths below.

When the water cleared all signs of the tragedy had vanished, but we still gazed into the silence, until the toot of the pinnace broke the spell.

"By the Holy M," said the bowman nervously, "that was a close call. Lucky for us the razorback came along or we might be making a trip to Davy Jones' locker, cuddled in the fond embrace of that slimy squid."

"Those eyes!" said another with a shiver, "I reckon I'll see them in my sleep until I die!"

Would you like to know what the intrepid Mr. Bootes said to Madam S. on the terrace of her house in Valparaiso? This is what he told her:

"I have never seen the gigantic cuttle at close quarters. The portions that have come under my notice have been vomited from the stomach of a dying cachalot—or sperm whale. Some of these parts have been at least twenty-five or thirty feet in length, having at the extreme end a series of discs, some a foot to eighteen inches in diameter. One of these specimens had, I remember, in addition to the adhering apparatus, a set of claws round the inside edge of the suckers, resembling the claws of a great crayfish. This I think proves that there are more than one species of *Sepia octopodia*, but I think they are all classified scientifically as Mollusca, which seems to include all soft-bodied animals in the sea. Then again the gigantic squid possesses, besides the eight branch legs or feelers, two tremendous tentacles. On some of these I have seen as many as six suckers, of all sizes, ranging from great basin-like receptacles to tiny ones, only about an inch in diameter. At the base—where they join the body—they are sometimes so thick that a man could scarcely embrace them, so you see madam, this creature is indeed tremendous."

And how did Madam S. respond to such a learned discourse? "Placing one hand over her eyes as though to shut out the awful picture my words had created in her mind," she said, "To students of marine natural history, your calling, Mr. Hedger [somehow, she was calling him by his middle name] offers unique fields of observation." Unfazed by the lady's sensitivity, Bootes then described a battle between a giant cuttlefish and a whale, and repeated "Captain N"'s account of the time his ship was seized by a great cuttlefish and the crew "spent two hours cutting away the arms and tentacles of the beast, which clawed at any and every thing within reach. He described how the some of the arms were fully twenty feet long and contained suckers which pulsated. The claw-like formations around the outer edges opened and shut long after they were severed from the body, suggesting something of its enormous vitality."

Many people assume that if *anyone* could see a giant squid (and get the science correct), it would be Jacques Cousteau, with his thousands of hours exploring the depths of the world's oceans. And yes, in his *Octopus and Squid: The Soft Intelligence* (written with Philippe Diolé), Cousteau records such a sighting:

> When I had reached 800 feet, I saw, through a porthole, a very large cephalopod, only a few yards from the minisub, watching the vehicle as it moved slowly past. I could not take my eyes from that mass of flesh, though it seemed not at all disturbed by the presence of the minisub. It was an unearthly sight, at once astonishing and terrifying. Was it sleeping? Or thinking? Or merely watching? I had no idea. It was there, nonetheless, enormous, alive, its huge eyes fixed on me. Then, suddenly, it was gone. I did not even see it move, though I am sure that an animal of that size is able to move with extreme rapidity by means of water jets from its funnel. The impression it made was one of size and power. I can understand how formidable a giant squid must be.

This may indeed be a description of *Architeuthis,* but it is so vague that one wonders whether it really happened or if Cousteau invented it. The location is not given (only that it was "during one of *Calypso*'s expeditions in the Indian Ocean"), nor is the date. All we can deduce is that it occurred before 1973, the publication date of the book. Other than "watching the vehicle as it moved slowly past," what was the animal doing? Did it move tail- or tentacles-first? What color was it? For that matter, what color were its "huge eyes"? How did an animal with one eye on each side of its cylindrical head look at Cousteau with what he described as "those great unblinking *eyes*"? (Italics mine.) And what does the last sentence mean? Because of its "size and power," did Cousteau understand how formidable *an opponent* a giant squid might be, or, having seen an animal that was a very large squid but not *Architeuthis,* did he wonder what a *real* giant squid might be like? At no point in the description can you find the words *I saw a giant squid;* it is all carefully worded inference, leaving it to the gullible reader to make the connection.*

Cousteau cross-references this discussion with "See *The Whale: Mighty Monarch of the Sea,* by Jacques-Yves Cousteau and Philippe Diolé." Published in 1972, the year before the octopus and squid book, *The Whale* reveals that also "in the Indian Ocean," *Calypso* found a "large white object" floating on the surface, which the crew retrieved and identified as "a piece of a giant squid's tail. The front part of it is torn, and it is covered with punctures similar to those inflicted by the teeth of a cachalot or a pilot whale."† They also retrieve "a piece of flesh shaped like a saucer, or rather like a plate. It is one of the squid's suction cups. Dr. François measures it and announces that its diameter is 24 inches. This, obviously was a 'small' giant squid. Its body probably measured between 8 and 10 feet—in addition to the large arms, of course." When they attempted to cook the pieces, they found that the tail was "so tough that we could not cut it," and "as for the suction cup, it was too horrible for us to describe. It was as though we had tried to make a meal out of a hunk of soft rubber."

There follows a brief discussion of the "fantastic Kraken," in which this sentence appears: "No man has ever seen an *Architeuthis* except as food not yet digested in the stomach of a sperm whale." Did he forget that *he* had seen one,

---

*In a 1996 letter to me, Kir Nesis, probably Russia's foremost teuthologist, wrote, "I have consulted with my colleagues having a rich experience in deep diving in manned submersibles. They say it is very difficult to observe large active oceanic squids in midwater. What they usually saw were quickly disappearing shadows, leaving only a large cloud of ink. . . . Thus I think J.-Y. Cousteau may see a very large cephalopod in midwater, but he could hardly identify and photograph it."

†On the page immediately following the description of the squid's tail is a glaring example of Cousteau's confused cetology. Even after he has found a piece with tooth marks, he declares, "The cachalot does not pulverize his food, nor does he chew. He does not even really bite. Instead he swallows his food whole, in a gulp." The latter interpretation is correct; squid taken from the stomachs of sperm whales show no evidence of having been bitten or chewed.

or did he just decide to leave it out of the whale book? In a discussion of eating (or trying to eat) pieces of *Architeuthis,* would he have forgotten that he had already described this "mass of flesh . . . enormous, alive, its huge eyes fixed on me"? And how could a man who claimed to have seen a living giant squid write that the "suction cup" they found measured 24 inches in diameter? (An ordinary dinner plate is 10 inches in diameter. A large garbage can lid is 20 inches in diameter. A New York City manhole cover is 27 inches in diameter.) I suggest that both events are fabrications: there was no sighting from a minisub, and there was certainly no attempt to cook a 24-inch-diameter suction cup, because there is no such thing as a 24-inch-diameter suction cup.*

Identified as "Architeuthis dux," this photograph appeared in a book called European Seashells in 1993. It is actually a photograph of Moroteuthis robustus *taken in northern Japanese waters as the squid was dying.*

---

*It is difficult to decide what to make of the mistakes, misspellings, and misrepresentations in the Cousteau book on cephalopods. If someone gets the facts wrong, the errors might be attributed to poor research, sloppy editing, or, in this case, imperfect translation. But what are we to make of such flagrant distortions? Since they are obviously not accidental, we must assume that the authors included this material intentionally, knowing that there was no possible way to back it up. Cousteau's reputation as a popularizer of marine natural history and conservation should have encouraged more, rather than less, dedication to accuracy.

In *European Seashells,* by Guido Poppe and Yoshihiro Goto, a book published in Germany in 1993, there appeared a photograph that purported to show a giant squid with a diver in the North Atlantic.* The diver in the photograph seems to be standing on the ground in shallow water, and there must have been another diver to take the picture. In a Japanese film made for television, there are scenes of a diver in very shallow water with a very large, very sick squid. It is clearly *Moroteuthis,* and comparison with the photograph in the Poppe book shows that it is the very same animal. As far as I know, there are no authenticated records of divers swimming with *Architeuthis*—in the North Atlantic or anywhere else. For a moment, I thought that some obscure photographer had captured the most elusive image in natural history. Fortunately for those who have devoted their lives to searching for *Architeuthis,* this was only an aberration, a case of mistaken identity.

---

*I tried, without much success, to track down the photograph and the photographer. Finally, having asked Guido Poppe for the name of the photographer, I learned that the picture was sold to the publishers (Hemmen in Wiesbaden) by IKAN, a photo agency in Frankfurt. They answered my query with this letter: "The giant squid (kalmar) in the Hemmen book was taken in southern Japan by a photographer represented by IKAN. But I doubt about *Architeuthis dux* as a valid identification. . . . I hope you are aware that the animal is not as big as it seems due to the wide-angle photography."

The
Models of
Architeuthis

Because it is one of the most spectacular animals on earth, the giant squid is often featured as the centerpiece of natural history museum exhibits. It is a natural for model makers, but since no one has ever seen a healthy *Architeuthis,* our knowledge of its morphology has come mostly from the examination of battered and broken specimens washed up on the beach. No photograph of a crumpled carcass can hope to do justice to a huge squid that, in its dark and cold natural habitat, swims almost weightlessly, with its great lidless eyes seeking the slightest glimmer in the blackness. And so far, there is no movie or video that shows a living giant squid, so how can one communicate the imposing size and exotic appearance of this fabulous creature, the world's largest invertebrate? The obvious answer is a model, and they have appeared in museums around the world since 1873.

In that year, when a giant squid washed ashore at Conception Bay, Newfoundland, its existence was communicated to Frank Buckland in England. Buckland (1826–1880) was a popular writer on hunting, fishing, and countless oddities of nature, as well as the founder and proprietor of the Economic Fisheries Museum in South Kensington. As soon as he read of this monster, it became clear to Buckland that he had to have the squid—or at least some sort of a facsimile—in his museum. According to official papers and photographs describing parts of the monster cuttlefish, the shape and length of body and tentacles were cut out in boards and hung in his museum, showing a sea monster with arms 35 feet long. Buckland's was the first attempt to show the public what a giant squid actually looked like.

In his 1879 discussion of the "gigantic squids *(Architeuthis)* and their allies," in which every Newfoundland specimen collected to date was discussed, A. E. Verrill of Yale University wrote:

> A nearly perfect specimen of a large squid, was found cast ashore in a severe gale, at Catalina, Trinity Bay, Newfoundland, September 24, 1877. It was living when found. It was exhibited for two or three days at St. John's, and subse-

quently was carried to New York, where it was purchased by Reiche & Brother for the New York Aquarium. There I had an opportunity to examine it, very soon after its arrival. . . . I am also indebted to the proprietors of the aquarium for some of the loose suckers. Other suckers from this specimen were sent to me from Newfoundland, by the Rev. M. Harvey.[*]

The squid's arrival in New York was documented by the *New-York Daily Tribune* for October 9, 1877:

THE AQUARIUM'S GREAT DEVIL-FISH.

By the steamer *Cortes,* from St. John's, Newfoundland, which reached her pier on Sunday, arrived the giant devil-fish, purchased by the Aquarium. On September 22, the monster was driven ashore at Catalina on the northern shore of Trinity Bay, by a severe storm, and was there captured alive, but it died a few moments afterward. It was taken to St. John's by the two fishermen who found it, and there Mr. Reiche, of the Aquarium made the valuable purchase. This great cuttle-fish (devil-fish is the more common name) is one of the very largest specimens ever taken. It measures forty feet and six inches from the point of the longest tentacle to the extremity of its tail. . . . It has not yet been determined whether to stuff the devil-fish or preserve it in alcohol. It was removed, without much difficulty, from the steamer to the Aquarium, yesterday afternoon.

On October 13, the *Daily Tribune* ran another story about the "devil-fish," to the effect that it was "shedding little bony rings, about large enough to slip on one's little finger. . . . They have formed the inner lining of some of the numerous suckers with which the arms of the animal are provided. Each ring has one edge serrated in points and it will be readily conceded that the close embrace of an octopus, thus armed, must be more penetrating than pleasant." The news-

---

*P. T. Barnum opened the first public aquarium in New York when he bought the Aquarial Gardens of Boston and then transferred its stock to his American Museum, located at Ann Street and Broadway in lower Manhattan. This institution burned to the ground in 1865, was rebuilt, and burned once again. With Charles Reiche and Brothers, entrepreneur W. C. Coup opened the first New York Aquarium at Broadway and Thirty-Fifth Street on October 11, 1876; this was the place that received the Trinity Bay giant squid. When the public was no longer interested, the aquarium closed down. The present New York Aquarium (now known as the Aquarium for Wildlife Conservation) opened in 1896 at Castle Garden at the southern tip of Manhattan Island, and is now located at Coney Island.

paper stories convey the sense of wonder that accompanied the arrival of this preternatural creature. It was virtually unknown to science in 1877 (Verrill had published the first description of *Architeuthis* only three years earlier), and it was certainly unknown to the public. It was as if someone had brought back a creature from outer space. On October 14, 1877, Verrill arrived in New York:

> Professor A. E. Verrill of Yale, visited the Aquarium this week to examine the great devil-fish, and compare the jaws with a pair of fish jaws in his possession, found several years ago in a whale's stomach.[*] He had been unable to account for a species of fish with such jaws as these, but in comparing them with the jaws of the devil-fish, the resemblance was striking. Professor Verrill said that the existence of a small, distinct species of these huge, ten-armed cephalopods has been clearly demonstrated, and that most of the specimens hitherto obtained had been taken in the Atlantic Ocean. . . . The form of the jaws of this specimen at the Aquarium constitute a powerful beak, looking something like that of a parrot or hawk, except that the upper jaw fits into the lower jaw, instead of the reverse, as in the beak of birds.

James Henry Emerton (1847–1931) illustrated the specimen "from nature" for Verrill's scientific paper, as he did many of the specimens that Verrill described. Verrill was evidently an accomplished draftsman himself, for he also did one of the drawings accompanying his 1879 discussion the "gigantic squids." Verrill wrote:

> The general form of this species is very well shown on Plate XX [Verrill's drawing]. This figure has been based upon the sketches and measurements made by me soon after the specimen was received in New York and before it had been "mounted." The head was, however, so badly injured that it could not be accurately figured, and this part is, therefore, to be regarded as a restoration, as nearly correct as could be made under the circumstances.

---

*In his 1879 discussion, Verrill wrote, "This specimen, consisting of both jaws, was presented to the Peabody Academy of Science, at Salem, Mass., by Captain N. E. Atwood, of Provincetown, Mass. It was taken from the stomach of a sperm whale, but the precise date and locality are not known."

In 1881, another Newfoundland squid arrived in New York, and again Verrill came down from New Haven to have a look at it. It had been collected at Portugal Cove, near St. John's, on November 10, and had been packed in ice and shipped to New York aboard the steamer *Catima*. It was exhibited at Mr. E. M. Worth's museum at 101 Bowery, where Verrill saw it. "Through the courtesy of Mr. Worth," wrote Verrill in 1882, "I have recently had an excellent opportunity for a satisfactory examination. Although a valuable specimen, it is not in all respects perfect, for the tips of all the sessile arms are broken off, and many of the suckers are gone; one-half of the caudal fin had apparently been lost, and the wound healed some time before the creature's death; the eye-balls had been burst in handling, and the pen mostly destroyed." Verrill wrote that he was mostly interested in this specimen for its internal organs, but he must have used some of its external appearance in the construction of his models.

The first full-sized model of a giant squid was made by J. H. Emerton under the direction of A. E. Verrill in 1883, and was displayed in the Peabody Museum, Yale College.

Using the two Newfoundland specimens for reference, Verrill and Emerton created the models for Yale and Harvard—the first full-sized *Architeuthis* models ever made. Verrill had studied under Louis Agassiz at Harvard from 1859 to 1862, and even though he was employed by Yale College when he made the model in 1877 (and thereafter until his retirement in 1907), Harvard also received a cast of this fabulous creature.

Emerton was a self-trained artist who specialized in natural history subjects, particularly invertebrates. (His passion was spiders, which might explain his interest in multilegged cephalopods.) In 1880, he was appointed assistant to Verrill, who was in the process of describing the giant squid that had been washing ashore in Newfoundland. Between 1874 and 1882, the Yale zoologist published twenty-seven papers on cephalopods, a great majority of which described various specimens of *Architeuthis*, and James Emerton provided the detailed illustrations for many of them.

According to a 1989 discussion of Verrill by Richard Johnson, "the life-sized model squid over forty feet long . . . was designed by Verrill but constructed by James H. Emerton of Salem, Massachusetts. . . . They were made of papier-mâché, rubber, and any other materials at hand. Three copies of the squid were made and two of the octopus. Only one model remains on exhibit [in 1989], that of a huge reddish octopus suspended from the Coral Room of the Museum of Comparative Zoology."

Although Emerton's article describes the making of the octopus model, we can safely assume that the squid was made in the same manner. Both were made of papier-mâché, cast in plaster molds:

> When the moulds were dry, the upper casts were made in them by methods used by Mr. Palmer for models of large fishes and cetacea. The moulds having been greased, paper soaked in paste was laid on it, and pressed and rubbed with the hands until it fitted close to the surface of the mould, and the edges of the pieces of paper adhered together. When the first layer of paper was nearly dry, another was pasted over it; and, if the strength of the model required it, other layers were added. . . . The model weighs about seventy pounds, and is stiff and strong enough for ordinary handling, and only liable to be broken by a fall or sudden blow."

In Emerton's obituary in the entomological magazine *Psyche*, he is credited with having written sixty-one papers, all on spiders. Nathan Banks, who compiled the obituary, apparently did not think that Emerton's discussion of cephalopod model-building was worthy of inclusion, inasmuch as it is not listed.

One of the earliest models of Architeuthis *was shown in the Museum of Comparative Zoology at Harvard University around the turn of the century. It no longer exists.*

Harvard's giant squid model deteriorated with time, and was eventually removed from exhibition. It was donated to the New England Aquarium in Boston around 1972, but the old model was in such poor condition that it was eventually discarded. Today, only the photographs attest to the dominating presence of the 40-foot-long giant squid model in the Cambridge museum. (When a real *Architeuthis* washed ashore at Plum Island, Massachusetts, in February 1980, it was first displayed at the aquarium, and then returned to the Museum of Comparative Zoology for study. The carcass is now in the Smithsonian's National Museum of Natural History, where, because they no longer have a full-sized model, a 12-foot-long juvenile *Architeuthis* is the centerpiece of the giant squid exhibit.)

Emerton's original model for the New Haven *Architeuthis* fared as badly as its Ivy League rival. After it had deteriorated to the point where it was no longer exhibitable, it was relegated to an empty coal cellar, and then discarded around 1964. But Yale rallied; in 1966, they built another one. In the Peabody Museum's *Discovery* magazine, the new model was announced thus:

> Upon entering the Hall of Invertebrates visitors will soon see suspended from the ceiling a giant squid, the largest known invertebrate, with a 12-foot body from which grope eight long arms and two 29-foot tentacles, all studded with hundreds of sharp-toothed suckers. Its iridescent green eyes measuring about ten inches in diameter had to be specially constructed out of plexiglass, as no commercial firms manufacture artificial eyes this size for any purpose. Anatomically, this reproduction is as nearly authentic as possible.

The 1966 Yale model was also based on descriptions by Verrill, but of a different specimen: one caught in Logy Bay, Newfoundland, in 1873. "More recent information was obtained from Frederick Aldrich of the Marine Biological Laboratory at Memorial University, St. John's, who studied squid found in Conception Bay," continued the article. "Mr. Townshend and Edward C. Migdalski, former Chief Preparator of Fishes of Peabody Museum, spent some time in Newfoundland last summer, making drawings and photographing preserved parts and gathering all available information to supplement the material at Yale." The article concluded:

> A small clay sculpture was then made as a working model. The reproduction is made of styrofoam with steel bracing and is covered on the outside with a layer of fiberglass which will be painted the squid's normal reddish hue. In order to depict the arms in motion which results in a more dynamic reproduction, more recently developed plastics, such as polyurethane foam were used. Heretofore models of giant squids elsewhere have represented them in almost rigid postures.

Ten years after Frank Buckland's wooden cutout was shown at his museum in South Kensington, another faux giant squid arrived in London. The U.S. Fisheries Commission had mounted a huge exhibit (officially known as the United States Fish Exhibition) for the International Fisheries Exhibition, also staged in South Kensington, from May 12 to November 1, 1883.* High over the cases showing the fishes that Americans caught and the equipment they caught them with was a model of *Architeuthis*. We can identify the origins of this model (and the giant octopus model that accompanied it) from a reference in an article by Emerton that appeared in the journal *Science* in 1883. The article is primarily about the making of the giant octopus model, but in it he says,

> While working on the models of the large Newfoundland squid (Architeuthis princeps) for the Yale and Harvard museums, it was proposed that I should also model the large octopus of the west coast of America. Nothing was done upon it, however, until the past winter, while preparations were being made by the U.S. fish-commission for the International fishery exhibition in

At the International Fisheries Exhibition in London in 1883, the giant squid model made for the Smithsonian by Verrill and Emerton hovers over the U.S. display.

---

*In *Nature* for July 26, 1883, an anonymous author wrote: "The American exhibit has the advantage of being the actual permanent collection of the National Museum of Washington, which has come into existence under the combined auspices of the United States Fisheries Commission and the Smithsonian Institution. The whole collection is not here, but we have a considerable part of it. For example an admirable series of coloured casts of the fishes of American waters, lifesize replicas of the gigantic Octopus and Architeuthis, a complete series of the Crayfishes of North America and of the edible crustacea generally, and samples of the more remarkable forms of life obtained in deep-sea exploration off the American coast."

London. For this exhibition, Mr. William Palmer, one of the modellers of the National museum in Washington, was sent to New Haven to make a copy of the Architeuthis model; and while this was in progress, plans for the Octopus were often discussed, and finally arrangements were made for him to remain in New Haven, to assist in making an Octopus model and a paper cast for the fishery exhibition.

In the catalog of the London exhibition, under the heading "Mollusca Cephalopoda: Squids and Cuttles," the giant squid model is described thus:

> *Architeuthis princeps,* Verrill. Giant Squid. Coast of Newfoundland and adjacent waters. Model made by Mr. J. H. Emerton, from measurements and descriptions of a Squid thrown ashore at Catalina, Trinity Bay, Newfoundland, September 24, 1877. Principal dimensions: Length of body, 8 feet; length of head, 1½ feet; length of tentacles, 30 feet; length of 1st pair of arms, 8½ feet; length of 2nd pair of arms, 9½ feet; length of 3rd and 4th pair of arms, 11 feet; greatest diameter of body, 2½ feet.

In his discussion of various cephalopods in the "illustrated and descriptive record" of the exhibition, Frederick Whymper wrote:

> If however, any lingering doubt existed as to the power of some of these creatures it must have been dissipated by a visit to the American Court of the Fisheries Exhibition. Two models, from the United States National Museum, constructed from the remains of actual examples, and the measurements of accurate scientific observers, were there to be seen hanging from the roof, the first representing the Octopus or Devil-fish *(Octopus punctatus),* and the second the Giant Squid *(Architeuthis princeps).* . . . The model of the giant squid was made from the remains, measurements, &c., of one thrown ashore at Catalina, Trinity Bay, Newfoundland, September 24, 1877.

Henry Lee, naturalist of the Brighton Aquarium, wrote several of the "handbooks" that were published in conjunction with the International Fisheries Exhibition. In one of them, *Sea Monsters Unmasked,* Lee described the process whereby the squid was collected and shipped to New York:

It was alive when first seen, but died soon after the ebbing of the tide, and was left high and dry upon the beach. Two fishermen took possession of it, and the whole settlement gathered to gaze in astonishment at the monster. Formerly it would have been converted into manure, or cut up as food for dogs, but, thanks to the diffusion of intelligence, there were some persons in Catalina who knew the importance of preserving such a Rarity, and who advised the fishermen to take it to St. John's. After being exhibited there for two days, it was packed in a half-ton of ice in readiness for transmission to Professor Verrill, in the hope that it would be placed in the Peabody or Smithsonian Museum; but at the last moment its owners violated their agreement, and sold it to a higher bidder. The final purchase was made for the New York Aquarium, where it arrived on the 7th of October, immersed in methylated spirit in a large glass tank.

Although the catalog credits Emerton with making the Washington model, in his own article (mentioned above), he says that "Mr. William Palmer, one of the modellers of the National museum in Washington, was sent to New Haven to make a copy of the Architeuthis model," so it appears that Palmer copied Emerton's Yale model. (It is clear from Emerton's discussion that Palmer worked with him on the octopus model.)

After crossing the Atlantic in both directions, the squid was finally installed in the Marine Invertebrates Gallery in the West Hall of the National Museum. It resided in the Smithsonian until it was removed from the museum and stored in the Butler Building, a Smithsonian storage facility in Silver Spring, Maryland. In 1992, the building was hit by a tornado and the roof torn off, resulting in severe flooding that turned the papier-mâché model into papier-mâché soup.

Some time before 1895, Emerton tried to sell a giant squid model to the American Museum of Natural History in New York, but his offer was refused. Eventually, a company in Rochester, called Ward's Natural Science Establishment, would take over the manufacture and sale of the models, and place them in such institutions as the California Academy of Sciences in San Francisco, the American Museum of Natural History in New York, the Field Museum in Chicago, the Milwaukee Public Museum, the Natural History Museum in London, and the Oceanographic Museum in Monaco.

Ward's Natural Science Establishment was founded by Henry Augustus Ward (1834–1906), who wanted to provide natural science materials for educational purposes. He began as a collector of minerals and fossils but soon enlarged the scope of his enterprise to include a broad spectrum of natural objects, from shells and skeletons to meteorites and mammoths. (One of his most popular items was a gigantic model of a mammoth, made of papier-mâché, with fake fur and plaster tusks. He was also responsible, with the help of a young taxidermist named Carl Akeley, for the mounting of P. T. Barnum's famous elephant, Jumbo.) In 1861, upon the sale of one of his mineralogical "cabinets" to the University of Rochester, that institution made him professor of natural science, and he remained close to the university for most of his life.

In *Henry A. Ward: Museum Builder to America,* written by his grandson Roswell Ward, we read that Ward's provided materials for the collections of the Field Museum (to the tune of $100,000); the Museum of Comparative Zoology at Harvard ($61,272); the University of Virginia ($51,006); Princeton ($32,405); the Coronado Beach Museum ($31,989); the American Museum of Natural History ($24,594), the University of Rochester ($22,769), "and seventy-four more colleges, universities and museums."

In the Rush Rhees Library of the University of Rochester, where the Ward's archives are housed, there is a letter from Emerton to Henry Ward, dated October 14, 1887, informing Ward that he would be arriving in Rochester the next day. It is likely that Emerton sold his drawings to Ward, for in Ward's 1892 catalog we find the following item being offered:

Genus *ARCHITEUTHIS,* STEENSTRUP, 1857.

ARCHITEUTHIS PRINCEPS, Verrill. Papier-mâché model.

This is one of the largest of the squids and measures 40½ feet in length, including the tentacular arms. The model is life size and was made according to measurements. The original specimen was captured on the coast of Newfoundland in 1877.

Ward's sold an enormous amount of material to the World's Columbian Exposition, held in Chicago in 1893 to celebrate the four hundredth anniversary of Columbus's discovery of the New World (accounting for Chicago's position at the top of the Ward's list), including the first pair of giant cephalopod models.

One of Ward's Architeuthis *models outdoors in Rochester(?), New York. These 40-foot-long models were not casually assembled, so this is probably a photograph that Ward's had taken as a sales device.*

After a visit to the exposition, Marshall Field was persuaded to donate one million dollars to create the Field Columbian Museum, which opened in 1894 at the building that had been the Palace of Fine Arts at Fifty-Seventh Street and Lake Shore Drive. (Marshall Field was the Chicago merchant who founded the store that bears his name, which was, from 1881 to 1906, the largest wholesale and retail dry-goods establishment in the world.) When Field died in 1906, he left some seven million dollars to the museum, which erected a new building and renamed it the Field Museum of Natural History. (It is now known simply as the Field Museum.)

From the 1894 accession records of the Field Columbian Museum, we learn that the collection included the jaws of a sperm whale, a skeleton of a giraffe, a pair of moose, an alligator, a collection of land shells, a collection of limpets, a collection of penguins, numerous other zoological specimens, as well as "1 model of octopus" and "1 model of squid." (The octopus and the squid were valued at five hundred dollars apiece.) According to the 1895 guide to the Field Columbian Museum, "Suspended over the table cases containing the invertebrates (shells, corals, crabs, etc.), was a life size model of a large squid. The original was found off the coast of Newfoundland in 1876."

In 1930, just across Lake Shore Drive from the Field, the Shedd Aquarium opened—an obvious location for the exhibition of a giant squid. But these models were rare and difficult to make, and over the years, the Field Museum's

Architeuthis was exhibited first at one institution and then at the other. In the spring of 1975, *Architeuthis chicagoensis* was moved to the Shedd, where it remained on exhibit until it was returned to the Field Museum in 1993—exactly a century after it had arrived in Chicago. At that time, according to Assistant Curator of Invertebrates Janet Voight, some minor gouges were repaired, and it was repainted "with an iridescent paint that gives it a better depth of color—maroon-bluish-reddish—depending upon the angle from which it is seen."

Henry Ward sold one of his first giant squid models to the California Academy of Sciences in San Francisco sometime before 1895. It was destined to be seen for only a few short years, for the earthquake and fire that ravaged the city did not spare the model—or the museum. On April 17, 1906, all of the collections and displays were destroyed, with the exception of the type specimens that were rescued as the building burned, and a few mammoth tusks that were only singed. Presumably, along with the remnants of the squid model, Ward's mineral and geological collections are now part of the jumbled stratigraphy of San Francisco.

Henry Ward also sold models to the American Museum of Natural History in New York. He began his sales pitch with an 1895 letter to Morris Jesup, president of the museum, in which he explained in detail why he was trying to sell him a model squid:

> I venture to write to you of one or two matters which may possibly have interest for you. . . . One piece . . . I have often been on the point of offering to the American Museum. This is the *Architeuthis princeps* or giant cuttle-fish from the mid-Atlantic, which was cast ashore ten years ago on the Newfoundland coast. There its remains were copied (molded) by men sent from the Smithsonian Institution by Prof. Baird. The work was overseen by Professor Verrill of Yale College; and three casts were made, which went to the Smithsonian, to Yale, and to Agassiz' museum at Cambridge, where they hang to-day, the admiration of visitors.
>
> The creature is 42 feet long. His body (cigar-shaped) is about 12 ft. long and 5 ft. in greatest diameter. From one end of the body grow his arms: eight of them very stout and about 12 ft. long, and two of them slender and 32 ft. long. These arms are provided with suckers—some 2,000 in number—by which the

strange animal fastens to its prey. . . . Professor Whitfield, who saw it here lately, has more than once said it should be in your great museum; hanging (it weighs but 500 lbs.) high over your collection of shells (it is, you know, a mollusc), or in some other suitable place.

I bought the molds of the *Architeuthis* some years ago, and have made four casts of it. One of these is in the museum of the Cal. Acad. of Sciences, another in the Field Columbian Museum at Chicago, a third in my own museum in Rochester, and the fourth is this one here. The two which I have sold were paid me $600 each. Now I am loaded down with this one here; and I so dislike to have to take it home that I have decided to offer it to you for the final sum of $450, if you will take it immediately. For this sum I will deliver it at your museum and there (my men helping me) mount it to cords furnished to me, pendant from the ceiling. The object can be seen here for another week, when you might have it inspected, seeing its absolute perfection in every way. Professor Whitfield saw it ten days ago and knows all about it.

On September 14, 1895, Professor Whitfield (the museum's curator of geology) sent a letter to Jesup in which he wrote, "The *Architeuthis princeps* model would be exceedingly desirable in our collection. I have mentioned this desiderata several times as the 'Giant-Squid' and recommended its purchase. The price Prof. W[ard] offers is much less than Mr. Emerton offered to make one for some years when the models were first made. It should also be accompanied by a model of the 'Giant Octopus' which was made at the same time."

Ward's importunements were successful; the American Museum's *Annual Report* for 1895 records that the Department of Marine Invertebrates acquired—"by purchase"—"1 Model of the Giant Squid (*Architeuthis princeps* Ver.), from Newfoundland," and "1 Model of the Giant Octopus (*Octopus punctatus* Gabb.) from California." A ledger entry confirms that the cost of them both was $750.

Ward's model at the American Museum has been exhibited in various locations in the museum over the past century: the Shell and Coral Hall, the Hall of Ocean Life, and the Invertebrate Hall. As of this writing, it has been dismantled (pardon the teuthid pun) and removed from exhibition prior to its reappearance as part of the new Hall of the Diversity of Life. When it reappears—along with the octopus—it will be 105 years old.*

If one were to assume that the presence of a giant squid model in Wisconsin, a state not known for its proximity to the ocean, was a result of the arrival of H. A. Ward's younger son, Henry Levi Ward, who became director of the Milwaukee Public Museum in 1902 and served until 1920, one would be wrong. The model was purchased by museum director Henry Nehrling, five years before Ward Jr. arrived on the scene. But not surprisingly, Henry L. Ward was involved in the sale of the squid (and octopus) models to Milwaukee. As was the case with New York and Chicago, Ward's sold the giant squid and the giant octopus as a set. On September 28, 1898, Henry Levi Ward wrote a letter to Henry Nehrling, director of the museum:

> We have today shipped by N.Y.C. freight prepaid a model of the squid *Architeuthis princeps* that my father evidently had conversed with you about. It is sent with the hopes that some of your wealthy friends may purchase it at its price of $450 and present it to the Museum. . . . If the squid is accepted perhaps we can send you the model of the Octopus later on. I am having one made now.

Ward's had provided the services of a Mr. Preston to the Milwaukee museum to assemble and hang the cephalopod models, paying him twenty-four dollars a week in addition to the fifteen dollars a week that the museum paid for his food and lodging. In March, Frank Ward, Henry Sr.'s brother, wrote:

> Our last bill was January 15th. Mr. Preston worked for four full weeks after this, leaving Milwaukee February 12th. This makes a month's salary that you owe us for. This will clear up the matter, except in case you do not keep the Squid and Octopus, you will then owe us fifty dollars a month additional for the last three months. Can you have this matter of the Squid and Octopus decided upon at the next meeting of your board?

---

*There is another life-sized giant squid model in the American Museum of Natural History. In the Hall of Ocean Life, a diorama shows a battle between *Architeuthis* and a sperm whale, but it is so dark that it is difficult to see what is going on. The whale's head and the squid's body and arms are sculpted in the round, but the squid's long tentacles are drawn in chalk on the back wall of the diorama.

In 1911, the Shell Gallery of the Natural History Museum in London boasted a life-sized model of a giant squid, made by Ward's of Rochester and installed in 1906. It was destroyed during the bombing of London in 1940.

The board finally decided to keep the models, for on July 21, 1899, Henry L. Ward wrote to Henry Nehrling: "We are in receipt of your esteemed favor of the 19th, enclosing your warrant for $800. . . . I am pleased that the Museum has decided to keep these two models as they certainly will go a long way towards interesting the public in your work."[*]

Recently refurbished, the Milwaukee squid model now hangs in a new exhibit called "A Sense of Wonder." For this installation, Wendy Chistensen-Senk of the museum's taxidermy department repainted it a "more passionate reddish maroon" instead of the battleship gray it had been for many years. (The octopus model, however, has not survived. It was cut in half and "repositioned" in the "Living Oceans" exhibit, where it now resides.)

The Natural History Museum in London has had two life-sized *Architeuthis* models, the first of which was exhibited in the Shell Gallery but was destroyed during the bombing of London in 1940. When I first saw the photographs of this model, I did not know who made it, but I thought it looked remarkably like those

---

[*]Floyd Easterman, the Milwaukee museum's curator of taxidermy, provided the Nehrling-Ward correspondence and also supplied a confirming date for the model. A close examination of one of its arms revealed a fragment of newspaper used in the papier-mâché laminations that is clearly dated 1897.

currently visible in New York, Chicago, and Milwaukee. According to the museum's archivist John Thackray, Ward's sold the usual pair (squid and octopus) to the British Natural History Museum in 1906: "The squid cost $450 and the octopus $350." Under the supervision of Director E. Ray Lankester, these models were installed in 1907.

At the turn of this century, Prince Albert-Honoré-Charles Grimaldi (1848–1922) established an oceanographic museum in the little principality of Monaco, perched on the Côte d'Azur of the northern Mediterranean. Work was begun on the museum in 1899, and it was formally opened in 1910. A guidebook to the museum, published around 1916, describes the model of *Architeuthis* that hung in the west hall:

> A life-sized model of a huge cuttlefish, so placed that it appears to be swimming in the water, hangs from the ceiling. It measures over 43 feet, including the tentacles, which alone measure 33 feet each.

The squid model (and its ubiquitous partner, the octopus) was sold to the museum for $450 in 1913 by Ward's Natural Science Establishment. The *calmar géant* was on exhibition in the Hall of Applied Oceanography from 1913 to 1995, but because the hall is now being refurbished, the model has been temporarily removed. (Henry A. Ward died in an automobile accident in Buffalo in 1906, but as we have seen in the case of the Milwaukee models, his brother Frank—who became president upon his death—and his son Henry Levi Ward were actively involved in sales.)

The *Architeuthis* models made by Emerton and Verrill, and later by Ward's Natural Science Establishment, were all patterned after the Newfoundland specimen of 1877, and possibly the one obtained in 1881 as well. The last of these models was the one made for the Oceanographic Museum of Monaco in 1913. There followed a long hiatus in the manufacture of giant squid models, but eventually, several other institutions decided that they, too, needed a replica of the world's largest and most elusive invertebrate.

The U.S. easily wins the title for having the most giant squid models, even though it is barely in the running for actual specimens washed ashore.* (There

---

*Devotees of paranormal phenomena might say that the reason the United States has done so poorly is because its name does not begin with an *N*. In descending order, the locations with the most giant squid beachings, sightings, and captures are New Zealand, Newfoundland, and Norway.

are a couple of unconfirmed records for California and Oregon, a definite one for Massachusetts, and a singleton found floating in the Gulf of Mexico.) Of the original total of nine models, Emerton made three and Ward's made six. The Emerton-Verrill models are all extinct, while the New York, Chicago, and Milwaukee models are still viable. The Ward's model in Monaco is still extant, but it is no longer on exhibition. The San Francisco specimen perished in an earthquake, and the pre–World War II British museum model was destroyed by the Luftwaffe.

The second British version of *Architeuthis* was constructed in 1974 by John Coppinger of the exhibition department of the Natural History Museum at their model-making studios in Cricklewood. (The museum was originally called the British Museum [Natural History], but it is now known as the Natural History Museum, or simply NHM.) With Malcolm Clarke as consultant, the model was constructed of fiberglass, with carbon fiber reinforcements for the tentacles. The new model is 11 meters long, or just over 36 feet. The molds—or *moulds* as they call them in Britain—for this model are now in the National Museum of Scotland in Edinburgh, because, according to David Heppell, curator of mollusca there, the NHM sold them as part of its recent "financial downsizing."

Using the same "moulds," Paul Howard, curator of the Yorkshire Museum, made a fiber-resin *Architeuthis* model, approximately 35 feet long. But this was not simply a case of wanting to have the world's most spectacular invertebrate on display; Yorkshire knows the giant squid firsthand. On January 14, 1933, W. J. Clarke received word that a 17-foot, 5-inch giant squid carcass had washed ashore (see pages 96–97). It was shipped to the Natural History Museum in London, where it was examined by G. C. Robson and pronounced to be a new species, which was named *Architeuthis clarkei*, after the man through whose efforts the specimen had been secured.

A glance at a map of Europe will show that the northern shores and island outposts of the British Isles and the southwest coast of Norway are on the periphery of the North Sea, and even though we don't know why such squid strand— or what they are doing in the North Sea in the first place—*Architeuthis* occasionally appears on these shores. To date, there have been fourteen recorded strandings of *Architeuthis* on the shores of Scotland (which includes the Shetlands and the Hebrides), and one in Yorkshire.

In the Haus der Natur in Salzburg, Austria, the writhing Riesenkalmar model easily dominates the exhibit hall.

Ireland, too, has seen a number of giant squid, so a model in the Ulster Museum in Belfast is not unexpected. One of the earliest known specimens (described on pages 66–67) was recorded from Dingle Bay, County Kerry, in 1673. During the decade from 1870 to 1880, when so many specimens were appearing on the barren beaches of Newfoundland, there were two Irish records. The first was caught at sea by three fishermen near Boffin Island, off the Connemara coast, on April 25, 1875. "The animal was found basking on the surface," wrote A. G. More (1875), "and was attacked by the fishermen who could not bear to think that so much good bait would be lost." In the end, they secured only the tentacles because the massive body "was too unwieldy, and was allowed to sink." The longest tentacle was 30 feet long, and the entire animal was estimated to have been 47 feet in length. Five years later, another specimen washed ashore at Kilkee, County Clare. Based on the Kilkee squid, model maker Sam Anderson of the Ulster Museum sculpted a model of polyurethane foam, and then made a smaller version for the aquarium in Dingle.

By and large, giant squid models appear in countries where giant squid appear, but there are some places—Austria, for example, with no coastline whatsoever—where the stranding of a giant squid would be most unlikely.

In Salzburg, in addition to the house where Mozart was born, there is the Haus der Natur. The "house of nature" exhibits many rare and exotic creatures, including dinosaurs of various kinds, and in the Great Hall we can see a life-sized

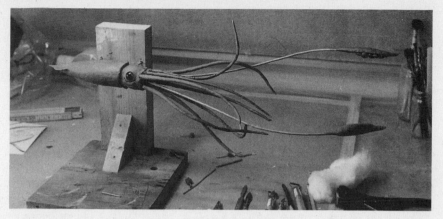

A miniature model used in the creation of the life-sized Architeuthis at the Zoological Museum of the University of Copenhagen.

model of the *Riesenkalmar.* According to the museum's description the model is 11.5 meters long (37.7 feet), and was sculpted by artist Wolfgang Grassberger from photographs.

The first giant squid officially recorded was one found floating in the Øresund, between Malmö and Copenhagen, in 1545. The model in the Zoological Museum of the University of Copenhagen was sculpted by Harry Hjortaa, and is based on the model in Salzburg, but the actual measurements are mostly those of a specimen that stranded in Ranheim (east of Trondheim), Norway, on October 2, 1954. Because of space limitations the two tentacles were shortened about 1.5 meters (4.9 feet) each to avoid their overlapping the tail of the sperm whale. The official opening of the Ocean Hall took place in November 1983 and was attended by Prince Henrik, Queen Margrethe, Crown Prince Fredrik, and Prince Joachim—surely the most illustrious assemblage of royalty ever to view a model of a giant squid.

Norway has a long, rich history of *kjempeblekksprut* strandings, and this fabled creature is well represented in Norwegian museums. The model at the Museum of Natural History in Trondheim is integrated into the ceiling, leaving only the arms and tentacles and the lower half of the body visible. One of the long tentacles reaches out and is curled around a model of a porbeagle shark, a unique touch. The Trondheim *Architeuthis* was modeled after a specimen now in their collection that stranded at Hemne on September 28, 1896. Created by Andreas

Esaissen and Emil Friden, the body is a cast, while the arms are carefully sewn fabric with cloth suckers attached. It was first exhibited in 1953, shortly before another specimen stranded at Ranheim on October 10, 1954. For about a week, the Trondheim museum had two giants on exhibit at the same time, but the Ranheim specimen was displayed in an unpreserved state, and was soon removed from exhibition and immersed in formalin.

The giant squid in the Bergen Museum was acquired in a rather more dramatic fashion. Joakim Lerøen was on the way to his boat at Austerheim, near Bergen, early one November morning in 1915, when he saw a gigantic squid swimming to and fro in a narrow bay. He gaffed the squid and dragged it ashore. (The details of this encounter can be found on pages 94–95.) The museum in Bergen was contacted the next day, and the carcass was collected shortly afterward. A plaster cast was made of the squid, and it was put on exhibit early in 1916. (Of all the models discussed in this chapter, this is the only one that is a cast rather than a "sculpture" based on measurements taken of one or more specimens.) The dimensions of the squid were: length of body to base of arms: 44.85 inches; length of first arm: 69.42 inches; length of longest tentacle: 18.2 feet. The arms of the squid were stretched out when the model was made, but the tentacles were arranged so that they curved back toward the tail fins. (At its full length, the Bergen squid was almost 24 feet long.)

From 1916 to 1983, the cast was standing upright in a corner on the museum's second floor, towering imposingly over the glass cabinets containing smaller invertebrates. Then, for the next few years, while the museum underwent reorganization, the public did not have access to this exhibit. Recently, the cast was relocated to the first floor, where it was mounted horizontally, high on a wall, behind a whale skeleton that hangs from the ceiling. In a letter to me, Professor Endre Willassen, curator of invertebrates at the Bergen Museum, wrote, "As you probably gather, I am not happy with the assignment of *Architeuthis* to a minor role among the other giants of the ocean."

Like Norway, Scotland, and Ireland, Newfoundland has a long history of giant squid landings. (The only place where there have been more *Architeuthis* records is New Zealand, but its list of specimens includes those caught in offshore trawling operations during the 1980s and 1990s.) The first identified Newfoundland specimen washed ashore at Lamaline in 1870, and since then, some thirty-six others have been recorded (Aldrich 1991). The specimen that A. E. Verrill used as a basis for his original models came from Catalina on Trinity Bay, but Dildo, seventy-five miles to the south, was the scene of another well-documented stranding in 1933.

When the citizens of Dildo wanted to create a tourist attraction, what more enticing subject could they think of than *Architeuthis*? Gerald Smith based the first endemic Newfoundland model on the animal that had been collected in 1933, relying on Nancy Frost's 1934 scientific paper, and also the eyewitness descriptions of twelve-year-old Reuben Reid and his friend Richard Gosse, who found the carcass floating in the bay. Smith and seven other Newfoundland craftsmen made their model out of fiberglass, and it is now the star attraction of the newly built (1996) Dildo Interpretation Centre.

What of New Zealand, the world leader in giant squid? There used to be a model in the museum in Wellington (officially known as Museum of New Zealand Te Papa Tongarewa), though Bruce Marshall, collections manager at the museum, wrote to me, "We had a life-size fiberglass *Architeuthis* model in our former marine gallery, but it was scrapped after the gallery was decommissioned . . . no space has been allocated for it in the new museum, and there are no plans to make another one."

At the Museum d'Histoire Naturelle on the rue Cuvier in Paris, a giant squid hovers over the Grande Galerie de l'Évolution. The polyester resin and fiberglass model is 10 meters (33 feet) in total length, painted in opalescent

Bob and Gail Cassilly made this model for the St. Louis Zoo and delivered it on a flatbed truck. It is the largest of all the giant squid models, and the only one not in a museum.

white with pink gradations. Although there are no suckers on the arms or the tentacular clubs, Jean-Philippe Maréchal of the museum informs me that they are planned for the near future. It was made by a Belgian firm in 1994 under the direction of Renata Boucher-Rodoni of the museum's Department of Marine Invertebrates, and was said to have been based on published illustrations, such those included in Roper and Boss's 1982 *Scientific American* article, and the 1991 study by Roeleveld and Lipinski of giant squid in South African waters.

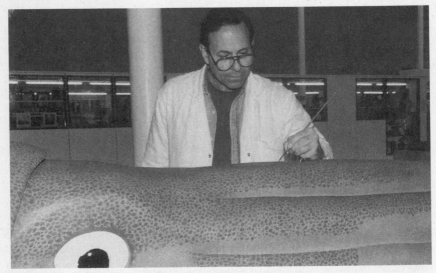

In June 1997, when the National Museum of Scotland in Edinburgh decided to hang a giant squid model, they invited the author of this book to paint it.

A couple of new models have appeared in the United States as well. In 1984, under the direction of resident teuthologist Eric Hochberg, the Santa Barbara Museum of Natural History commissioned a giant squid model by professional model maker Carl Gage of Sierra Madre, California. Probably unique among all the squid models, this one was based on measurements of a *Japanese* specimen (there are no giant squid models in Japan, as far as I can tell) that is known as *Architeuthis japonica*. Made of glass fiber and polyester resin, it is 23 feet long from the tip of the longest tentacle to the point of the tail. If you wanted one of Gage's models for your museum (or for your swimming pool), it would run you a cool fifty thousand dollars.

Another model has been set up in St. Louis, Missouri, about as far from an ocean as you can get in the United States. Designed and made by Bob and Gail Cassilly, the model squid at the St. Louis Zoo is the only one in America that is not in a museum. Bob Cassilly consulted with Clyde Roper by telephone, but based a lot of his design on smaller squids that he bought in a local fish market. It was sculpted out of plywood, fitted onto a steel-pipe framework, and then sprayed with urethane foam, which was augmented with auto-body putty. It was then carved to shape, sanded, and painted by Gail Cassilly. The suckers were cast from halved cherry tomatoes, cherries, and grapes, with a nail driven through to make the stalk. In November 1988 the 52-foot-long, 600-pound model— probably the largest model on exhibit anywhere—was delivered to the zoo on a flatbed truck.

The National Museum of Scotland in Edinburgh has specimens of *Architeuthis* (one of which stranded at North Berwick in 1977), but, until recently, it had no model. In May 1997 the museum obtained a cast of the model in the Natural History Museum in London, and through the good offices of David Heppell, curator of mollusca, I was invited to Scotland to paint it. It had originally been spattered with a sort of reddish yellow paint, so I decided to delineate the chromatophores, those little sacs in the skin that can be enlarged or reduced by the animal to change its color. I painted what seemed like an infinite number of little dots, half of them Venetian red and half carmine, to give the model an overall reddish appearance, and I repainted the eyes so they were no longer black with a gold-rimmed pupil. To the catalog of *Architeuthis* models, Edinburgh's may now be added.

A model of a giant squid is not terribly complicated. As noted by many who have tried to make one, its body has a simple fusiform shape with a pair of tail fins at one end and a crown of arms and tentacles at the other. The eyes are big enough (and their actual appearance doubtful enough) so that any self-respecting taxidermist or glassworker can make a reasonable facsimile. (Suckers seem to be a matter of national choice, or, as the French say, *chacun à son goût*.) Why then should the story of these models turn out to be so convoluted? Perhaps it has to do with the very nature of the subject: we are not on familiar terms with this animal, the largest known invertebrate. *Architeuthis* certainly is an elusive creature. Its occasional appearance on various beaches around the world has provided hardly more than a glimpse of its majestic and intimidating appearance, and hauling it out of the water in a trawl does no justice to it either. Papier-mâché or fiberglass models have given us a sense of its size and shape, but they have not captured its mystery and vitality. The spirit of *Architeuthis* may well be uncapturable; at least no museum has even come close to this fabulous creature—the largest living animal that has never been seen alive.

Conclusion

quids are not part of our world, not elements of our consciousness. They are endowed with features—hooks, claws, suction cups, lights, beaks, a mucous coating, multiple appendages—that we rarely encounter in the more familiar terrestrial creatures. They live out of sight, underwater, at depths we cannot plumb, in numbers we cannot imagine. Their strength, their competence, and their predominance in their oceanic habitat have inspired some authors to classify them as an alternative form of intelligence on the planet. Their unfamiliar shape, with a cluster of arms at one end, eyes in the middle, and a tail at the other end, has only added to the impression that they are alien creatures from an unknown world— which is exactly what they are.

They are so little known that filmmakers and novelists have to devote too much time and space to an explanation of how the animal *works* before they can get it to attack anybody. (All that business of slimy skin, eight grasping arms, talons, a parrotlike beak, and so on.) The maximum known size of the giant squid has also contributed substantially to its image; an animal that can reach a length of 60 feet is already intimidating, and if it happens to have eight squirmy arms, two feeding tentacles, gigantic unblinking eyes, and a gnashing beak, it becomes the stuff of nightmares. It lives in a world that is hardly known to us: the black, icy reaches of the world's deepest oceans, where it cruises (we are not even sure whether it swims arms- or tail-first) in search of its prey—whatever that might be.

Is *Architeuthis* a solitary hunter, or does it congregate in schools? The former is somehow more comforting, since the vision of a pack of 60-foot-long gigantic squid is truly terrifying. (In a 1967 article in the British magazine *Animals,* Frederick Aldrich wrote, "Giant squids are not as rare as was once believed. A school of 60 has been sighted off the coast of Newfoundland, and only last autumn we heard of a surface battle between a giant squid and a whale." In his subsequent publications, Aldrich never again mentioned the school of sixty giant squid, so it is likely that he learned it was not accurately reported.) Indeed, the vision of just one of these creatures is horrifying enough. That may be because we are so unfamiliar with anything that is so big. One of our great fears is that creatures that are terrifying in appearance at a small scale—insects for example—will be suddenly enlarged so that we can see their eight eyes, or their horrible pincer jaws, or their ominously waving antennae. Directors make horror movies about giant ants, flies, or roaches, but they don't have to exaggerate the size of *Architeuthis*—it is *already* a monster.

Since Aristotle wrote about the *teuthos,* people have been fascinated by giant cephalopods. From Olaus Magnus to the Bishops Pontoppidan and Egede and the Reverend Harvey, right up to Arthur C. Clarke, Michael Crichton, and Peter Benchley, there has been a uninterrupted chain of *Architeuthis* enthusiasts drawn from the ranks of historians, writers, and even artists.

To the scientists who study the various forms of squids, they are among the most fascinating animals on earth. The scientist is captivated by the very attributes that repel or disgust the layman. Gilbert Voss of the University of Miami served as adviser and mentor to a generation of cephalopod biologists, and wrote seventy-six papers on squids and octopuses for scientific journals. He was born in 1918 and died in 1989; upon his death, the *Bulletin of Marine Science,* of which he had been the editor, published a memorial volume as a tribute to him. In their introduction to this 1991 volume, Clyde Roper and Michael Sweeney wrote,

> It seems certain that Professor Gilbert L. Voss was responsible for stimulating and influencing a new wave, a resurgence, of research on cephalopods that has been sustained and increased over his 40-year career. So important is his influence on the resurgence of cephalopod systematic research, hardly a paper has been published in cephalopod systematics and zoogeography, and even other aspects of cephalopod biology and fisheries, that doesn't cite at least one paper by Voss. This has been the case for nearly 4 decades and it will continue to be so for many decades into the 21st century.

Gil Voss was not a fan of *Architeuthis.* He believed it was a weak, slow-swimming creature, "with arms and tentacles that [are] quite flabby and easily torn loose from the body." But he loved *Dosidicus,* which he said (in a 1959 article entitled "Hunting Sea Monsters") were "rulers of their domain, of quite a different nature from *Architeuthis*." "Their bullet-shaped bodies," he wrote, "are heavy and strong, with powerful jet funnels and large fins. Their arms and tentacles are massive and strong and with their beaks they can bite oars and boat hooks in two and eat giant tunas to the bone in minutes."

In the same article, Voss wrote that hypothetical giant ommastrephid squids "would be among the most powerful fighting machines the marine world has ever produced, and there is no reason to believe they cannot exist." And when he wrote about squids in general for *National Geographic* ("Squids: Jet-Powered Torpedoes of the Deep") in 1967, he said, "Few marine animals possess traits as fascinating, or flaunt such jewel-box beauty as these transparent mollusks. . . . The cephalopods include the swiftest swimmers in the sea, except perhaps for some game fishes, and exhibit behavior bordering on active intelligence."

Because they are such fascinating creatures, giant squid have attracted the attention of a dedicated fraternity of specialists who have devoted a substantial portion of their professional careers to the study of *Architeuthis*. From the first, there were not very many specimens to work with, and when Japetus Steenstrup erected the genus that incorporates his name, he based it on a set of jaws and some historical records that had been published three hundred years earlier. Steenstrup was in touch with A. E. Verrill of Yale, who, as the beneficiary of the influx of giant squid on the beaches of Newfoundland, described and cataloged them with an enthusiasm verging on obsession. Born in 1927, one year after Verrill died, Frederick Aldrich was a New Jersey native who took his Ph.D. at Rutgers, worked at the Academy of Natural Sciences in Philadelphia, and remained at Memorial University of Newfoundland from 1961 until his death in 1991. He claimed to have examined more giant squid than anyone in history.

Glen Loates is a Canadian wildlife artist, probably best known for his painstakingly detailed paintings of North American mammals and birds. Whenever possible, he works from life, but so far, his lifelong attraction to the giant squid has not allowed for on-site sketching. Loates befriended Frederick Aldrich and traveled from his home near Toronto to Newfoundland to observe Aldrich's pickled specimens. He carefully drew those, perfecting his familiarity with the minute anatomical details of the giant squid. Then he created a series of drawings and paintings of *Architeuthis* in action—often in a struggle with a sperm

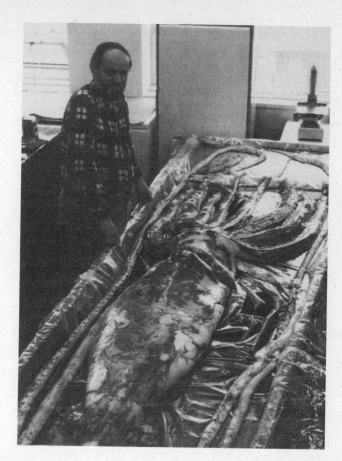

Frederick Aldrich in his laboratory at Memorial University in Newfoundland, with one of the fifteen giant squid carcasses he examined.

whale—that are certainly the most accurate and exciting depictions of the monster ever drawn. In the *Journal of Cephalopod Biology*, Frederick Aldrich introduced the Glen Loates portfolio by writing, "His enthusiasm for giant squid dates to his first reading, as a child, of Jules Verne's *Twenty Thousand Leagues Under the Sea*, and at the age of 16 he first attempted to commit the architeuthid to paint and canvas. In his Toronto home he has assembled a cephalopod library of which many a teuthologist could well be proud."

In a 1974 interview in the *Los Angeles Times*, Aldrich said, "I've got to get me a live giant. . . . We will use lights, chopped-up shark as bait in a basket, a big jig. Divers will walk the beast into the boat." Fifteen years later, with shark scientist Eugenie Clark and National Geographic photographer Emory Kristof, Loates and Aldrich were participants in the Beebe Project, a submersible effort dedicated to finding the giant squid. The first lowerings of the submersible *Pisces VI* were made off Bermuda, a region not known for giant squid. (The only reports of *Architeuthis* in Bermuda waters had come from monster hunters.) When this proved fruitless, the expedition was relocated to Bonavista Bay, Newfoundland, and Fred Aldrich joined them there. In 1990, Aldrich descended in the submersible to a depth of eight hundred feet hoping to see his favorite animal, but no giant squid appeared. In hope of attracting a giant squid, they lowered a gigantic squid jig, painted bright red and outfitted with numerous hooks, but this didn't work either. Despite his lifelong commitment to *Architeuthis,* Fred Aldrich died without having seen a living giant squid.

Of course there are teuthologists who study *Architeuthis* in passing; if one (or more) shows up on your beach or in your laboratory, you are likely to become an instant convert. Among those who have published descriptions of giant squid (always dead) are Eric Hochberg of California, Kir Nesis of Moscow, Jean Cadenat of France, Nancy Frost of Canada, G. C. Robson of England, Martina Roeleveld of Cape Town, R. K. Dell and Steve O'Shea of New Zealand, David Heppell and Peter Boyle of Scotland, Joyce Allan, Wolfgang Zeidler, Mark Norman, and C. C. Lu of Australia, France Staub of Mauritius, and any number of Scandinavian scientists who found themselves the often unwilling custodians of a huge, flaccid, ammonia-scented carcass, usually with the eyes missing.

Scientists today have the benefit of the work done by their predecessors: Steenstrup was *Architeuthis*'s first champion; though he wrote prose that even his own biographers called "characteristically involved and often rambling," one can identify his enthusiasm for the giant mollusks. Without A. E. Verrill's voluminous publications and meticulous descriptions, Frederick Aldrich would not have been able to create a basis for his studies of Newfoundland squid, and Clyde Roper, who led the 1997 expedition to New Zealand , would have had considerably less information on the history and habits of *Architeuthis.* Roper, born in 1937 and a student of Gil Voss's, is now curator of invertebrates at the Smithsonian's National Museum of Natural History and, like his predecessors, is an apostle of the church of *Architeuthis.* All of these men pursued other interests (only Roper was trained as a teuthologist; he wrote his doctoral thesis on *Bathyteuthis,* an Antarctic species); Japetus Steenstrup, although he did not have a university degree, was a polymath who published 239 papers on paleontology, archaeology, and history, as well as zoology. For these professionals, as well as for us amateurs who desire only to fathom the oceans' most fantastic and mysterious predator, it was the fabled *Architeuthis*—gigantic, lurking, hovering, unblinking, unseen—that captured their imaginations in an enduring tangle of powerful tentacles.

Long after I had completed this manuscript and turned it over to the publisher, I received a letter from a man who had seen me claiming on a television program that nobody had ever seen a living giant squid. "Not true," said Dennis Braun, now an operations analyst for a large defense contractor, and he continued:

"In 1969 (about February or March), just prior to my going to Vietnam, I was a 19-year-old Marine Corporal on an amphibious assault training exercise to Vieques Island, near Puerto Rico. We had made an encampment on one of the Island's beaches; stayed for about two weeks and were preparing to return to Morehead City, North Carolina (via the U.S.S. *Francis Marion*), when I and two of my fellow Marines saw something quite extraordinary in the water alongside our ship, which was at anchor.

We had come back aboard ship on the first day of making preparations for re-embarkation, because we had large quantities of equipment to be stowed and there was work to be done in securing this gear in the ship's holds as it was craned aboard from the shuttling landing craft. During lulls in the activity there was lots of idle time on deck, and the sailors were prepared for this by having their own fishing tackle; so we watched them fish.

The water in that area is very clear, having a light sandy bottom, and with no aquatic growth visible around the ship, you could see straight to the bottom. It was like looking into a swimming pool. The weather was sunny and pleasant and the water very calm. Large yellowtail hung around the ship in schools of 20 or 30, and that's what the sailors were mostly interested in.

We saw sharks and a few other odds and ends, and it was about mid-day on the second or third day that I and two friends were leaning on the port-side railing looking South toward Vieques Island (about half a mile or so distant). Someone from the other side of the ship suddenly came over and announced loudly something like, "Hey, you ought to see the size of the barracuda this guy just hooked!" So naturally just about everybody rushed over to that side to watch the battle. Engrossed in our conversation, and figuring there'd be time anyway, my friends and I had stayed behind momentarily — continuing to talk — when I shifted the focus of my vision from the sunlight dancing on the water, to down toward the bottom near the ship. What I saw was astonishing!! There, in full view on the sandy bottom, a huge squid had come to rest! We were amazed as we all three looked on at the thing.

Apparently it had just decided to lounge alongside the ship, parallel to it, but still in the sun, and you could clearly see even its eyes as it lay there on the sand; its head to our left, and its tentacles, fanned slightly, extending to our right. The length and girth were truly astounding, and I can only guess as to its actual dimensions, but it was MUCH larger than the one shown laying dead in the TV episode. It just laid quietly very near the ship, being maybe twenty or so feet out from directly below us, and pretty much centered on where we stood (along its length). I would estimate that to scan its entire bodylength from my position above the water (about 30 ft. up) and by looking straight down, I would have to rotate my eyes at least 30 degrees to the left and 30 degrees to the right.

Its body looked big enough in diameter that I couldn't begin to get my arms around it (not that I'd want to), but maybe halfway. I have little doubt that the thing could have probably taken down a sport-fishing boat, if it chose to, just based on its sheer size and weight! The coloration looked pretty normal for what you'd expect in a squid; brownish; darker than the surrounding sand, but almost like it had taken on some of that lighter color.

There was absolutely no doubt at all as to what it was. The water must have been at least fifty or more feet deep for the ship to be there, yet, in its clarity — and from our place on the deck — which was as I would again guess, maybe about thirty feet above the water's surface — we had a very penetrating vantage point. I realize that, based on my assessment of the distance from me, and the angles described above, it was probably at least 100 ft. in length.

It just laid there unmoving and seeming to be looking back up at us as we chattered about it and what to do. I remember suggesting that maybe we should go up to the bridge and tell someone. But no, we reasoned in our naiveté, they probably already knew it was there because of Sonar and whatever other sensing equipment they might have — our not realizing that Sonar wouldn't likely even be switched-on when at anchor. Besides, as young Marines, we were all kind of intimidated about approaching officers about anything, much less something like this!

Altogether, I'd say we had watched it for ten or more minutes. Its outline and features were unmistakable and the time we observed it allowed us to scrutinize and compare our observations pretty much without the benefit of imaginative excess.

I know there are those who would dismiss what I claim to have seen — as drifting seaweed, or the shadows of clouds or the like, but I promise you — if you were to take a highly detailed rubber squid 10 ft. long and sink it in the deep end of a swimming pool — it was just as clear as that. There were no other structural features around it, and Disney couldn't have done a better job of making it look real.

Vieques Island is immediately south of the Puerto Rico Trench, which, at 27,500 feet, is the deepest part of the Atlantic Ocean. There have been no records whatsoever of giant squid from Puerto Rico — or Cuba, or Hispaniola, or any other islands that lie along this vast chasm. The Bahamas, almost 1,000 miles away, have produced two carcasses, both found floating on the surface: one was by Steenstrup's Captain Hygom in 1855, and the other was brought to Gil Voss in 1948.

I spoke to Dennis Braun, and he was totally innocent of the controversy that surrounds the giant squid. In fact, although it was very large, he had no reason to assume that the creature he saw was in any way unusual. In the years since 1969, he told me, he has told his story to many people, but evidently he spoke to no one who knew enough about giant squid to tell him that his was a most unusual experience, perhaps unique.

If it is true (and it really was *Architeuthis*), then it adds an entirely new dimension to the mystery of the giant squid. Up to now, I have dismissed stories of sightings of giant squid that do not fall within the accepted maximum known length of 55 feet. But what if Bernard Heuvelmans or Arne Grønningsaeter or Fred Aldrich or J. D. Starkey are correct, and there really *are* 100-footers out there? And how about Arthur C. Clarke's contention that the largest ones are probably not the ones that washed up? (I'm only willing to include the 100-footers in this amendment; a 150- or 200-foot-long monster would have a very difficult time finding enough food to keep it going, and is still, I believe, in the realm of fantasy.)

Have I been too quick to dismiss the unmeasured giants and too unwilling to modify the statement that "nobody has ever seen a living giant squid"? What if Dennis Braun's story is true? And when I wrote that giant squid are probably not schooling animals, I hadn't seen a little story that appeared in the January 1998 issue of the *Marine Observer*, published in London. In it, a ship's officer captain named C. A. McDowall describes an incident that occurred "some years ago . . . in the Arabian Sea at night:

> . . . we were visited by a large school of giant squid — I think. They just rose out of the deep to look at us, about 200 of them. There were babies the size of a bucket, and adults, the biggest having bodies 3-4m long with two long tentacles about another 6m long. We lowered the loading ramp to get a good look, and the Captain's granddaughter took photographs — which probably didn't come out because of the very bright lights and the creatures being in shadow. The eyes were very large, bigger than a dinner plate, but the most remarkable thing was the colour. The top of the head was red, like a Ferrari, and the tentacles were white covered with red spots which made them look pink. Where the red back joined the white area around the eyes, there was a pattern of interlocking spots. The crew tried to catch the babies but once hooked they broke free, and that individual could not be hooked again, which was interesting. They stayed for about an hour and a half, and then slowly sank from view. I mention this because I have heard on several occasions that 'no-one has ever seen the Giant Squid' but I do not think this can be so. . . . Has anyone else seen these creatures? It was interesting also because it is an area where there are female sperm whales.

Perhaps, instead of categorically clinging to the closed system I have worked so hard to establish, I ought to end this book with an open mind. Maybe there are bigger giant squid out there, and maybe people have actually seen them. That they haven't been measured may not be all that important anyway. Can we dismiss every account that doesn't fall within our established guidelines? Will the inclusion of such stories detract from the verisimilitude of this book? On balance, I think it is better to include these stories than to ignore them.

We know we are supposed to believe in it, but still we doubt. Can there really be a 60-foot-long creature with unblinking dinner-plate eyes in the unknown vastness of the icy depths? The existence of *Architeuthis* only confirms our fears and inadequacies; despite our puny efforts to capture or understand it, the monster perdures. What will happen if someone finds it or takes its picture? It will lose some of its mystery, and, in a sense, we will be poorer for having been deprived of the anticipation of finding it. Often the realization of a long-held goal proves less fulfilling than the hungry waiting. In *The Log from the Sea of Cortez,* John Steinbeck wrote, "Men really need sea-monsters in their personal oceans. . . . For the ocean, deep and black in the depths, is like the low dark levels of our minds in which the dream symbols incubate and sometimes rise up to sight like the Old Man of the Sea. . . . An ocean without its unnamed monsters would be like a completely dreamless sleep." We need to find the giant squid, but we also need not to find it.

Acknowledgments

In my search for *Architeuthis,* many people have contributed, in laboratories, museums, and libraries around the world. My introduction to teuthid biology was aided and abetted by John Arnold, Ron O'Dor, Jennifer Hoar, Bernd Budelmann, Eric Hochberg, Clyde Roper, David Heppell, Martina Roeleveld, Kir Nesis, Uwe Piatkowski, Wolfgang Zeidler, and Malcolm Clarke. John Arnold retired to his home and lab in Massachusetts before the arrival of the E-mail blizzard, but I have pestered almost everyone else with cyber-questions, and I am still amazed at the speed and ease with which the answers arrive in New York from such far-flung locations as Adelaide, Wellington, Tokyo, Cape Town, Kiel, Copenhagen, Bergen, Edinburgh, Kaliningrad, Moscow, Santa Barbara, Milwaukee, New Haven, and Cambridge, Massachusetts. With Roger Hanlon of the Marine Biological Laboratory at Woods Hole, I have spent many productive hours discussing our mutual fascination with the behavior of cephalopods. (It was also convenient for me that he and John Messenger wrote a book on the subject.) Fran Hoskin, also of the MBL, patiently explained the biochemical aspects of squid enzymes to me. Colm Lordan of County Cork provided the information and the photographs of the trio of Irish architeuthids that were caught in 1995. When the exciting news of the Australian specimen with embedded spermatophores broke in 1997, I contacted C. C. Lu and Mark Norman, and asked them about the parts I didn't understand. Both of them were patient and more than willing to explain the intricacies of the bizarre sex life of the giant squid.

My friend Mary Petersen of the Zoological Museum of the University of Copenhagen provided me with material on Japetus Steenstrup that I was previously unaware of, greatly improving my (still limited) understanding of the earliest discoveries of *Architeuthis;* she also kept a watchful eye on my use of various Danish and Norwegian words. Mary introduced me to Jan Haugum, who effectively corrected my bumbling interpretations of Scandinavian languages, and sent me translations of various Danish works otherwise not available in English. Mary also put me in touch with Yuri Nekrutenko of Kiev, who translated many of the Russian papers for me.

For the even more problematical names of squids in Japanese, I am grateful to John Richard Bower of the Faculty of Fisheries, Hokkaido University, who walked me gently through the thickets of Japanese etymology and phonetics. I have had some very interesting conversations about the nature and behavior of the giant squid with Peter Benchley, even though we have agreed to disagree about the presence of "claws" on the tentacles of *Architeuthis.* I thank Glen Loates, who has generously permitted me to reproduce his wonderful illustrations of giant squid in action. Loren Coleman of the University of Maine shares my obsession with *Architeuthis,* and even though I know he wanted to write a book like this, when I told him I was doing it, he graciously opened his files to me.

Because there are no pictures of living giant squid, I decided to use photographs of models throughout this book to give a sense of what the creature looks like. I thought the search for models would be easy—just call up the museum and ask them to send me a photograph and the history of their model. Instead, it turned out to be a complicated detective story, with vanishing models, mysterious sculptors, lost records, and even a locked ledger that had to be opened by a locksmith. For their assistance on this case, I would like to thank the Squid Squad, who helped me track down various versions of *Architeuthis artificialis:* Nina Cummings, the Field Museum's photo librarian, led the midwestern division, while Mary DeJong of the American Museum of Natural History library delved deeply into the dusty archives to come up with missing records and photographs. Jim Atz, researcher and bibliographer extraordinaire, came to my rescue again, as he has done on so many previous occasions. For the information on the California Academy of Sciences' architeuthid earthquake victim, I am grateful to my friend (and occasional co-author) John McCosker; and for material on the Smithsonian's tornado casualty, Clyde Roper and Mike Sweeney. In Edinburgh, David Heppell, curator of mollusca at the National Museum of Scotland, provided invaluable assistance in his own backyard, and also directed me to Ulster and York. (He also asked me to come to Scotland to paint the Edinburgh model.) From Ulster, Angela Ross sent me data and the photograph, and from Yorkshire, Registrar Melanie Baldwin put me in touch with Paul Howard, who actually made the model. I learned about the models in the British Natural History Museum through the careful guidance of Archivist John Thackray and Assistant Librarian Paul Cooper, while the photographs were graciously provided by Lodvina Mascarenhas. Bill Muntz of Monash University in Australia put me on the trail of Frank Buckland. Mary Petersen, who was so helpful with the *Kjempeblekksprut,* arranged for the photographs from the Copenhagen Museum, and helped me find the Haus der Natur in Salzburg. When I finally found it, Dr. Norbert Winding sent me the photographs, and Claudia Kraehmer did the necessary translations. Endre Willassen of the Bergen Museum provided information on and the photographs of one of the two Norwegian models, and Torleif Holthe of Trondheim supplied the same for the other. Susan Otto and Floyd Easterman of the Milwaukee Public Museum helped me with their model and others. The Ivy League was represented by Ken Yellis, Michael Anderson, Eric Lazo-Wasem, and Barbara Narendra of Yale, and Jim McCarthy and Eva Jonas of Harvard's Museum of Comparative Zoology. Eric Hochberg of the Santa Barbara Museum of Natural History provided the details and the pictures of their model, and Carl Gage explained how he made it. At the St. Louis Zoo, information was provided by Liz Forrestal Reinus, and for pictures of the model they made, I thank Bob and Gail Cassilly. There is now a giant squid model in Newfoundland, and I found out about it from Lillian Simmons of the *Compass,* who also arranged for the photograph. I had written to C. C. Lu in Melbourne, asking him about models in Australia, but his reply came from Paris. Having taken early retirement from the Museum of Victoria, he was in France studying the cephalopod collections, and he put me in contact with Renata Boucher-Rodoni, who in turn arranged for Jean-Philippe Maréchal to send me the data and pictures of the Parisian *calmar géant.* When I thought I had all the models, I learned that there was yet another one at the Oceanographic Museum of Monaco; in the nick of time, Mauricette Hintzy told me about it, and then Christian Carpine provided the background information. For the history and correspondence of Ward's Natural Science Establishment (and photographs of the models before they were sold), I am especially grateful to Karl Kabelac of the Special Collections at the University of Rochester's Rush Rhees Library.

Bernard Heuvelmans, the doyen of cryptozoology, wrote a book in 1958 about "le kraken et le poulpe colossal" (translated in 1997 as *The Kraken and the Colossal Octopus*), in which he discussed many of the subjects included in this book. Some portions of his book appeared in English in an abridged form in his 1965 *In the Wake of the Sea-Serpents,* but the full English translation of *The Kraken* was not completed until 1996. Paul LeBlond, the translator, provided me with a prepublication version in early 1997. I admire Heuvelmans's scholarship and dedication to the subject, but I respectfully request his indulgence to disagree on many points.

For assistance and guidance—not to mention support and friendship—I will always be indebted to Nina Root, the librarian of the American Museum of Natural History.

I can produce almost anything in a hard copy, from the Reverend Moses Harvey's 1899 article about how he found the first of the Newfoundland giant squids, to Otto Nordgård's discussion (in Norwegian) of cephalopods observed in Trondheimsfjord in 1923, and Dr. Schwediawer's comments on *Ambergrise,* published in 1783. But there is a new resource in the land—actually, it's in the phone lines—and that is E-mail. Of course, given my proclivities, I print out and save all my E-mail messages, but now I have encountered a completely different problem: how do I acknowledge information that is only a personal communication or, even more difficult, a message distributed to a list-server like Deepsea, Eurosquid, or Mollusca? These are discussion groups on topics of particular interest, in this case squids and octopuses, particularly the abyssal forms. They allow one to post a message (for example, "does anyone know about the 47-foot *Architeuthis* that was collected by Gilbert Voss in 1957?") to dozens or hundreds of people simultaneously, and get answers almost immediately. (The answer— from Nancy Voss at the University of Miami, where the specimen was supposed to have been stored—is that only an arm was saved.) Dennis Braun first told me about the Vieques Island giant via E-mail, but I later talked to him on the telephone, so as to bring the tale into a more "responsible" format.

In those cases where I have received useful information through a Web site or from E-mail, I have tried to acknowledge authorship in a traditional fashion.

But the purpose of a bibliography is to allow another person to check a reference, or consult a particular work. How can anyone else find messages that were sent to me two years ago from Tasmania or Japan? I have tried to incorporate such information as effectively and honestly as possible, but some stray, undocumentable factoids may have crept into the manuscript. I accept full responsibility for any errors, cyber- or otherwise, that appear in this book.

I first met Steve O'Shea of New Zealand's National Institute of Water and Atmospheric Research (NIWA) through the Internet, but when he accompanied the *Architeuthis* specimen from Auckland to New York in June 1998, we exchanged much useful information, and he read and commented on the manuscript in an early form. Likewise, Clyde Roper read the entire manuscript, and made many suggestions and corrections that I hope have kept me from completely embarrassing myself.

It is not often that an author acknowledges himself, but observant readers will recognize that portions of this book appeared—sometimes verbatim—in my 1995 *Monsters of the Sea.* That book was written largely because of my fascination with the giant squid, and because I felt that I had said some things as well in *Monsters* as I could, I have repeated them here.

Bryan Oettel watched nervously as the manuscript grew larger and more complex, and cringed as I argued for more convoluted taxonomic interpretations and longer excerpts. Many writers regard editors as adversaries who want to remove what they have lovingly written, and all authors know that their prose is too valuable to sacrifice for crassly commercial objectives like length or cover price. But Bryan had the best interests of this book (and this writer) at heart, and if he had not been constantly vigilant, the book would have been twice as long and half as readable.

And once again, Stephanie has been there, watching patiently as I chased these elusive creatures all over the globe (usually in libraries or via the Internet) and tried to make sense out of the architeuthid mysteries. There are not many people who would be happy having a book about squid dedicated to them, but for Stephanie's loyalty and support, this book is for her.

Authenticated Giant Squid Sightings and Strandings

Because some of the specimens had tentacles that were lost or damaged, some of the overall lengths include tentacles, and some do not. (In one instance, it was *only* a tentacle that was found.) Because the arms are often damaged, a more reliable measurement is the mantle length, abbreviated as ML. In some cases, the original description was not seen by me, and I have had to use a secondary source as a reference.

DATE	LOCATION	SIZE	AUTHOR
1545	Malmö, Denmark	8.16'	Steenstrup (1854)[1]
1639	Thingøre Sand, Iceland	Unknown	Steenstrup (1849)
1673	Dingle Bay, Ireland	19'	More (1875)
1770	Jutland, Denmark	Unknown	Muus (1959)
1785	Grand Banks, Newfoundland	Unknown	Thomas (1795)
1790	Arnarnesvik, Iceland	39'	Steenstrup (1849)
1798	Denmark	Unknown	Packard (1873)
1802	off Tasmania	6–7'	Peron (1807)
1817	Atlantic Ocean	400 lbs.	Quoy & Gaimard (1824)
1853	Raabjerg, Denmark	(beak only)	Steenstrup (1857)

DATE	LOCATION	SIZE	AUTHOR
1855	Bahamas (Captain Hygom)	Unknown	Steenstrup (1857)
1855	Aalbækstrand, Denmark	Unknown	Muus (1959)
1860	Hillswick, Scotland	23'	Jeffreys (1869)
1861	*Alecton* (Canaries)	20–24'	Bouyer (1861)
1862	North Atlantic	Unknown	Crosse & Fischer (1862)
1870	Waimarama, New Zealand	15'+	Kirk (1880)
1870	Lamaline, Newfoundland	40'	Verrill (1879)
1871	Lamaline, Newfoundland	47'	Verrill (1879)
1871	Grand Banks, Newfoundland	15'+	Packard (1873)
1871	Wellington, New Zealand	16'	Dell (1952)
1872	Bonavista Bay, Newfoundland	46'	Verrill (1879)
1872	Coomb's Cove, Newfoundland	52'	Verrill (1879)
1873	Portugal Cove, Newfoundland	44'	Harvey (1874)
1873	Logy Bay, Newfoundland	32'	Murray (1874)
Unknown	Labrador	52'	Verrill (1879)
1874	Fortune Bay, Newfoundland	36'	Verrill (1879)
1874	Buøy, Norway	Unknown	Grieg (1933)
1874	*Strathowen* & *Pearl* (Indian Ocean)	Unknown	Lane (1974)
1875	Connemara, Ireland	30' tentacles	More (1875)
1875	St. Paul Island, Indian Ocean	Unknown	Velain (1877)
Unknown	Cape Sable, Newfoundland	43"	Verrill (1879)
1876	Hammer Cove, Newfoundland	Unknown	Verrill (1879)
1876	Cape Campbell, New Zealand	20'	Kirk (1880)
1877	Lance Cove, Newfoundland	44'	Verrill (1879)

DATE	LOCATION	SIZE	AUTHOR
1877	Catalina, Newfoundland	39.5'	Verrill (1879)
1878	Thimble Tickle, Newfoundland	55'	Verrill (1879)
1878	Three Arms, Newfoundland	31'	Verrill (1879)
1878	James's Cove, Newfoundland	40'(?)	Verrill (1879)
1879	Brigus, Newfoundland	8' arms	Verrill (1879)
1879	Lyall Bay, New Zealand	(beak only)	Kirk (1880)
1880	Tokyo Fish Market	Unknown	Hilgendorf (1880)
1880	Island Bay, New Zealand	55'	Kirk (1882)
1880	Tønsvik, Norway	Unknown	Grieg (1933)
1880	Kvaenangen, Norway	Unknown	Grieg (1933)
1880	Co. Clare, Ireland	Unknown	Ritchie (1918)
1880	Grand Banks, Newfoundland	66" tentacles (juvenile)	Verrill (1880)
1881	Portugal Cove, Newfoundland	21'	Verrill (1882)
1886	Cape Campbell, New Zealand	Unknown	Robson (1887)
1887	Lyall Bay, New Zealand	57'	Kirk (1888)
1895	Tokyo Bay	12.5'	Mitsukuri & Ikeda (1895)
1896	Hevnefjord, Norway	32' 2"	Brinkmann (1916)
1896	Hevnefjord, Norway	32'	Nordgård (1928)
1898	North of Bahamas	Unknown	Steenstrup (1898)
1902	*Michael Sars* (off Faeroes)	9.35'	Murray & Hjort (1912)
1903	Mjofjördur, Iceland	tentacle	Murray & Hjort (1912)
1909	Truro, Mass.	17'+	Blake (1909)
1911	Senjen, Norway	20' 4"	Grieg (1933)

DATE	LOCATION	SIZE	AUTHOR
1911	Monterey Bay, Calif.	30'	Berry (1912)
1912	Monterey Bay, Calif.	c. 500 lbs.	Berry (1914)
1912	Japan (fishing net)	Unknown	Pfeffer (1912)
1912	Smølen, Norway	28' 4"	Brinkmann (1916)
1914	Belmullet, Ireland	27' (in sperm whale)	Hamilton (1914)
1915	Bergen, Norway	23' 8"	Brinkmann (1916)
1916	Helgeland, Norway	Unknown	Grieg (1933)
1916	Hevnefjord, Norway	20' tentacles	Nordgård (1928)
1917	Skateraw, Scotland	Unknown	Ritchie (1920)
1918	Kilkel, Ireland	Unknown	Hardy (1956)
1918	Tokyo Fish Market	Unknown	Heuvelmans (1965)
1919	Øyvagen, Norway	Unknown	Nordgård (1923)
1920	Hebrides, Scotland	Unknown	Ritchie (1920)
1922	Caithness, Scotland	Unknown	Ritchie (1922)
1924	Bluff, New Zealand	16'	Dell (1952)
1924	Margate, Natal	Unknown	Heuvelmans (1965)
1927	Kalveidøy, Norway	24.6'	Grieg (1933)
1928	off Greenland (Godthaab Expedition)	jaws only	Muus (1962)
1928	Ranheim, Norway	46'	Nordgård (1928)
1930	East Lothian, Scotland	10.48' (no tentacles)	Stephen (1961)
1930	Kaikoura, New Zealand	41'	Dell (1952)
1930	Miura Peninsula, Japan	26' 6"	Tomilin (1967)

DATE	LOCATION	SIZE	AUTHOR
1930–33	Pacific between Hawaii and Samoa	Unknown	Grønningsaeter (1946)
1933	Scarborough, Yorkshire	17.5'	W. J. Clarke (1933)
1933	Dildo, Newfoundland	Unknown	Frost (1934)
1935	*Palombe* (Gulf of Gascogne)	26'	Cadenat (1936)
1935	Harbour Main, Newfoundland	20'	Frost (1936)
1937	Arbroath, Scotland	24.5'	Stephen (1937)
1937	Petone, New Zealand	22' tentacles	Dell (1952)
1939	Tromsö, Norway	42' 8"	Wood (1982)[2]
1945	Pahau River Mouth, New Zealand	Unknown	Dell (1952)
1946	Vike Bay, Norway	30'	Myklebust (1946)
1948	Wingan Inlet, Australia	28'	Allan (1948)
1949	Bay of Nigg, Scotland	19' 3"	Rae (1950)
1949	Shetland, Scotland	beak only	Stephen (1950)
1949	Hirtshals, Denmark	5' 9"	Muus (1959)
1950–55	Gulf of Mexico (Mississippi Delta)	2' ML; 9' TL	Voss (1956)
1952	Madeira	34'	Rees & Maul (1956)
1952	Carnoustie, Scotland	Unknown	Hardy (1956)
1952	Florida Keys	36" ML	Voss (1996)[3]
1954	Ranheim, Norway	30'	Knudsen (1956)
1955	Porto Pim, Azores	34' 5" (in sperm whale)	R. Clarke (1955)
1956	Makara, New Zealand	5.9' ML	Dell (1970)
1957	Aberdeenshire, Scotland	18.49'	Stephen (1961)
1958	Bahamas	47'	Voss (1967)

DATE	LOCATION	SIZE	AUTHOR
1958	Nordfjord, Norway	9.62'	Kjennerud (1958)
1961	Madeira	57 mm ML (juvenile)	Roper & Young (1972)
1961	King's Cove, Newfoundland	"small"	Aldrich (1968)
1963	off Chile	45 mm ML (juvenile)	Roper & Young (1972)
1964	Conche, Newfoundland	28.76'	Aldrich (1991)
1964	Chapel Arm, Newfoundland	4' ML	Aldrich (1991)
1965	Lance Cove, Newfoundland	16' tentacle	Aldrich (1991)
1965	Springdale, Newfoundland	19.82'	Aldrich (1991)
1966	Sweet Bay, Newfoundland	56" ML	Aldrich (1991)
1966	Wild Cove, Newfoundland	41" ML	Aldrich (1991)
1966	Eddies Cove, Newfoundland	poor condition	Aldrich (1991)
1969	East of Lake Worth, Fla.	Unknown	Voss (1996)[3]
1970	San Juan, Puerto Rico	Unknown	Voss (1996)[3]
1970	off St. Pierre, Newfoundland	Unknown	Aldrich (1991)
1971	Sunnyside, Newfoundland	69" ML	Aldrich (1991)
1972	off Durban	Unknown (in sperm whale)	Roeleveld & Lipinski (1991)
1972	off South Africa	Unknown (in blue shark)	Nigmatullin (1976)
1974	off South Africa (trawl)	Unknown	Pérez-Gándaras & Guerra (1989)
1974	Green Point, South Africa	14.82'	Roeleveld & Lipinski (1991)
1975	Bonavista, Newfoundland	51" ML	Aldrich (1991)
1975	Trondheimsfjord, Norway	26.10' TL	Holthe (1975)
1976	off South Africa	34.5'	Pérez-Gándaras & Guerra (1978)

DATE	LOCATION	SIZE	AUTHOR
1977	Firth of Forth (Scotland)	Unknown	Heppell (1977)
1977	Lance Cove, Newfoundland	Unknown	Aldrich (1991)
1978	Fort Lauderdale, Fla.	6.5" ML	Toll & Hess (1981)
1978	Cheynes Beach, Western Australia	Unknown	*Sea Frontiers*
1979	St. Brendan's, Newfoundland	60" ML	Aldrich (1991)
1979	Spanish trawler, Newfoundland	31' TL	Stephen, pers. comm. (1997)
1980	Southern California	tentacle only	Robison (1989)
1980	South Africa	2.65' ML	Pérez-Gándaras & Guerra (1989)
1980	off Oregon (trawl)	5.3' ML	Nesis *et al.* (1985)
1980	off California (trawl)	2.5' ML	Nesis *et al.* (1985)
1980	Plum Island, Mass.	30'	Roper (1982)
1980–81	East North Pacific (trawls)	17 specimens	Nesis *et al.* (1985)
1981	Hare Bay, Newfoundland	29.96'	Aldrich (1991)
1981	Orange River mouth (South Africa)	2.63'	Nesis *et al.* (1985)
1981	off Sydney	10.3 mm ML	Lu (1986)
1981	Vavilov Ridge (off Zaire) (trawl)	3.25' ML	Nesis *et al.* (1985)
1982	Bergen, Norway	33'	Brix (1983)
1982	Sandy Cove, Newfoundland	65" ML	Aldrich (1991)
1982	Sydney, Australia	16.5" ML	Jackson *et al.* (1991)
1982	Niigata Prefecture, Sea of Japan	13.05'	Honma *et al.* (1983)
1983–88	New Zealand waters	24 specimens	Gauldie, West, & Förch (1994)
1984	off Namibia	4' ML	Pérez-Gándaras & Guerra (1989)
1984	Aberdeen, Scotland	13' 9"	Boyle (1984)
1986	Aberdeen, Scotland	14'	Boyle (1986)

DATE	LOCATION	SIZE	AUTHOR
1986	Orange River mouth (South Africa)	15' 6"	Roeleveld & Lipinski (1991)
1987	off South Africa	2.6' ML	Pérez-Gándaras & Guerra (1989)
1989	off Brazil	58" ML	Arfelli *et al.* (1991)
1989	off Namibia	22.16'	Villanueva & Sánchez (1993)
1991	Soetwater, South Africa (off Cape Town)	13' 2"	Natal *Mercury* (1996)
1991	Cape Point (off Cape Town)	1.4 m ML	Roeleveld, (1996)
1992	Kommetjie (South Africa)	head only	Roeleveld, (1996)
1992	Cape Columbine (South Africa)	1.77 m ML	Roeleveld, (1996)
1993	Mauritius	4.5 m ML	Staub (1993)
1995	South Australia	30.3'	Zeidler (1996)
1995	Southwest Ireland	3 specimens	Lordan (1997)
1996	off Tasmania	32'	Norman & Lu (1997)
1996	off Tasmania	24'	Norman & Lu (1997)
1996	off Tasmania	7.8' ML	Norman & Lu (1997)
1996	Chatham Rise, New Zealand	26'	O'Shea (1996)
1996	Chatham Rise, New Zealand	13'	O'Shea (1996)
1996	New Zealand	22–26'	O'Shea (1996)
1996	Tottori, Japan	14.4' TL	*Japan Times* (1996)

NOTES

[1] Prior to 1648, the southernmost part of Sweden, known as the Skania region, belonged to Denmark. Therefore, the town of Malmö was Danish when the creature was found there in 1545. Muus (1959) writes that "the oldest Danish find originates from 1545, when a specimen was caught live near Malmö. From contemporary descriptions with accompanying woodcuts it appears that the animal was regarded as a 'soemunk.' Japetus Steenstrup delivered a lecture in 1854 with a strong suggestion that the 'soemunk' was an *Architeuthis*."

[2] In *The Guinness Book of Animal Facts and Feats,* Gerald Wood writes that "the largest known *Architeuthis* was one killed by fishermen near Tromsö on 10 October 1939, which measured over 13m [42.6'] and had 8.7m [28.5'] long tentacles." The reference for these figures is "Karl Basilier, personal communication." Other than Mr. Basilier's communication with Mr. Wood, there does not seem to be another reference to this specimen.

[3] In August 1996, Nancy Voss sent me the complete list of "architeuthids [whole or partial specimens] catalogued in the Rosenstiel School of Marine and Atmospheric Sciences (RSMAS) Marine Invertebrate Museum." Three of these do not appear in any published literature, and are listed here only on the basis of the existence of a partial specimen in the museum's collection.

References

Abrahamson, D. 1992. Elusive Behemoth [Giant Squid]. *Rodale's Scuba Diving* October 1992:106, 118.

Akimushkin, I. I. 1963. Cephalopods of the Seas of the U.S.S.R. *Izdatel'stvo Akademii Nauk SSSR.* Moscow. (Israel Program for Scientific Translation. Jerusalem. 1965.)

Aldrich, F. A. 1967. Newfoundland's Giant Squid. *Animals* 10(1):20–21.

————. 1968. The Distribution of Giant Squids in N. Atlantic and Particularly About the Shores of Newfoundland. *Sarsia* 34:393–98.

————. 1987. Moses and the Living Waters: Victorian Science in Newfoundland. In D. H. Steele, ed., *Early Science in Newfoundland and Labrador,* pp. 86–120. Sigma Chi, St. John's, Nfld.

————. 1989. A Portfolio of Drawings of *Architeuthis dux* by Glen Loates. *Jour. Ceph. Biol.* 1(1):87–93.

————. 1991. Some Aspects of the Systematics and Biology of Squid of the Genus *Architeuthis* Based on a Study of Specimens from Newfoundland Waters. *Bull. Mar. Sci.* 49(1-2):457–81.

Aldrich, F. A., and M. M. Aldrich. 1968. On Regeneration of the Tentacular Arm of the Giant Squid *Architeuthis dux* Steenstrup (Decapoda, Architeuthidae). *Canadian Jour. Zool.* 46:845–47.

Aldrich, F. A., and E. L. Brown. 1967. The Giant Squid in Newfoundland. *Nfld. Quarterly* 3:4–8.

Aldrovandi, U. 1613. *De piscibus libri V et de cetis liber unus.* J. C. Unterverius. Bologna.

Allan, J. 1948. A Rare Giant Squid. *Aust. Mus. Mag.* 9:306–08.

———. 1955. The Kraken—Legendary Terror of the Seas. *Aust. Mus. Mag.* 11:275–78.

Anderson, R. C. 1996. Records of *Moroteuthis robusta* (Cephalopoda: Onycho-teuthidae) in Puget Sound (Washington State, U.S.A.). *Of Sea and Shore* 19(2):111–13.

Anon. 1849. The Sea-Serpent. *Pictorial National Library* 3(19):263–8.

Anon. 1912. Large Squid on the Oregon Coast. *Nautilus* 25:23.

Anon. 1966. New Developments in Exhibit Areas [Peabody Museum, Yale University]. *Discovery* 1(2):45–46.

Anon. 1979. Squid Squib. *Pacific Search* 13(4):38.

Anon. 1980. Giant Squid. *Australian Fisheries* April 1980:27.

Arata, G. F. 1954. A Note on the Flying Behaviour of Certain Squids. *Nautilus* 68(1):1–3.

Arfelli, C. A., A. F. Amorim, and A. R. G. Tomás. 1991. First Record of a Giant Squid *Architeuthis* sp. Steenstrup, 1857 (Cephalopoda: Architeuthidae) in Brazilian Waters. *Bull. Inst. Pesca.* 18:83–88.

Aristotle. N.d. *Historia animalium.* Loeb Classical Library. Harvard University Press.

Arnold, J. M. 1984. Reproduction in Cephalopods. In *The Mollusca,* vol. 7, pp. 419–54. Academic Press.

———. 1991. Frederick Allen Aldrich, 1927–1991. *Jour. Ceph. Biol.* 2(1):77–78.

Ayling, T. 1982. *Collins Guide to the Sea Fishes of New Zealand.* William Collins.

Baker, A. de C. 1957a. Some Observations on Large Oceanic Squids. *Ann. Rep. Challenger Soc.* 3(9):34.

———. 1957b. Underwater Photographs in the Study of Oceanic Squid. *Deep-Sea Res.* 4:126–29.

———. 1960. Observations of Squid at the Surface in the N.E. Atlantic. *Deep-Sea Res.* 206–10.

Banks, J. 1783. An Account of Ambergrise, by Dr. Schwediawer; Presented by Sir Joseph Banks. *Phil. Trans. Roy. Soc. London* 73(1):226–41.

Banks, N. 1932. J. H. Emerton [Obituary]. *Psyche* 39(1-2):1–8.

Barber, L. 1980. *The Heyday of Natural History, 1820–1870.* Doubleday.

Barber, V. C. 1968. The Structure of Mollusc Statocysts, with Particular Reference to Cephalopods. *Symp. Zool. Soc. London* 23:37–62.

Bardarson, G. 1920. Om den marine molluskenfauna ved vestkystn af Island. *Kgl. Dansk. Vid. Selsk. Biol. Medd.* 2:1–139.

Barlow, J. J. 1977. Comparative Biochemistry of the Central Nervous System. In M. Nixon and J. B. Messenger, eds., *The Biology of Cephalopods,* pp. 325–46. Academic Press.

Bartsch, P. 1917. Pirates of the Deep—Stories of the Squid and Octopus. *Rep. Smithsonian Inst.* 1916:347–75.

———. 1931. *The Octopuses, Squids and Their Kin.* Smithsonian Scientific Series 10:321–56. Washington, D.C.

Bassler, R. S., C. E. Resser, W. L. Schmidt, and P. Bartsch. 1931. *Shelled Invertebrates of the Past and Present.* Vol. 10, Smithsonian Scientific Series. Washington, D.C.

Beale, T. 1835. *A Few Observations on the Natural History of the Sperm Whale.* London.

Beebe, W. 1931. A Round Trip to Davy Jones's Locker. *National Geographic* 59(6):653–78.

———. 1932a. The Depths of the Sea. *National Geographic* 61(1):5–68.

———. 1932b. Flying Squids. *Bull. N.Y. Zool. Soc.* 35(5):177–78.

———. 1932c. *Nonsuch: Land of Water.* Harcourt, Brace.

———. 1934a. A Half Mile Down. *National Geographic* 66(6):661–704.

———. 1934b. *Half Mile Down.* Harcourt, Brace.

Bel'kovich, V. M., and A. V. Yablokov. 1963. The Whale—an Ultrasonic Projector. *Yuchnyi Tekhnik* 3:76–77.

Belloc, G. 1954. Calmars géants. *Bull. trimestriel musée océanographique de Monaco* 32:11–12.

Belon, P. 1551. *L'histoire naturelle des étranges poissons marins avec la vraie peinture et description du Dauphin, & de plusiers autres de son espèce.* R. Chaudière.

———. 1555. *La nature et la diversité des poissons avec leurs pourtraicts, représenté au plus près du nature.* C. Estienne.

Belyayev, G. M. 1962. Rostra of Cephalopods in Oceanic Bottom Sediments. *Okeanologiia* 2(2):311–26.

Benchley, P. 1991. *Beast.* Random House.

Bennett, F. D. 1840. *Narrative of a Whaling Voyage Round the Globe, from the Year 1833 to 1836 . . . with an Account of Southern Whales, the Sperm Whale Fishery, and the Natural History of the Climates Visited.* Richard Bentley.

Bergman, B. 1982. Tentacled Encounters. *Equinox* 3(3):130.

Berry, S. S. 1909. Diagnoses of New Cephalopods from the Hawaiian Islands. *Proc. U.S. Nat. Mus.* 37:407–19.

———. 1910. A Review of the Cephalopods of Western North America. *Bull. U.S. Fish. Bur.* 30:267–336.

———. 1912a. The Cephalopoda of the Hawaiian Islands. *Bull. Bur. Fish. Wash.* 32:255–362.

———. 1912b. Note on the Occurrence of a Giant Squid off the California Coast. *Nautilus* 25:117–18.

———. 1914. Another Giant Squid in Monterey Bay. *Nautilus* 28(2):22–23.

———. 1916. Cephalopoda of the Kermadec Islands. *Proc. Acad. Nat. Sci. Phila.* 68:45–66.

———. 1920a. Light Production in Cephalopods, I. An Introductory Survey. *Biol. Bull.* 38(3):141–69.

———. 1920b. Light Production in Cephalopods, II. An Introductory Survey. *Biol. Bull.* 38(4):171–95.

Berzin, A. A. 1972. *The Sperm Whale.* Izdatel'stvo "Pischevaya Promyshlennost" Moskva 1971. (Israel Program for Scientific Translation 1972.)

Bidder, A. M. 1964. Feeding and Digestion in Cephalopods. In K. M. Wilbur and C. M. Yonge, eds., *Physiology of Mollusca,* vol. 2, pp. 97–124. Academic Press.

Bille, M. A. 1995. *Rumors of Existence.* Hancock House.

Blake, J. H. 1909. A Giant Squid. *Nautilus* 23:43–44, 83.

Boletzky, S. V. 1977. Post-Hatching Behaviour and Mode of Life in Cephalopods.

In M. Nixon and J. B. Messenger, eds., *The Biology of Cephalopods,* pp. 557–67. Academic Press.

Bompas, C. G. 1885. *Life of Frank Buckland.* Smith Elder & Co.

Boorstin, D. J. 1983. *The Discoverers.* Random House.

Bouyer, M. 1861. Poulpe géant observé entre Madere et Ténériffe. *Comptes rendus des séances de l'académie des sciences* (Paris).

Boycott, B. B. 1953. The Chromatophore System of Cephalopods. *Proc. Linn. Soc. London* 164:235–40.

Boyle, P. R. ed. 1983. *Cepahalopod Life Cycles, Vol. 1: Species Accounts.* Academic Press.

———. 1986. Report on a Specimen of *Architeuthis* Stranded near Aberdeen, Scotland. *Jour. of Molluscan Studies* 52:81–82.

Breland, O. 1953. Which Are the Biggest? *Natural History* 62:67–71.

Bright, M. 1989. *There Are Giants in the Sea.* Robson.

Brinkmann, A. 1916. Kjaempeblekkspruten (*Architeuthis dux* Stp.) i Bergens Museum. *Naturen* 40(6):175–82.

Brix, O. 1983. Giant Squid May Die When Exposed to Warm Water. *Nature* (London) 303:422–23.

Brix, O., A. Bårdgard, A. Cau, A. Colisimo, S. G. Condò, and B. Giardina. 1989. Oxygen-Binding Properties of Cephalopod Blood with Special Reference to Environmental Temperatures and Ecological Distribution. *Jour. Exp. Zool.* 252(1):34–42.

Broad, W. J. 1994. Squid Emerge as Smart, Elusive Hunters of Mid-Sea. *New York Times* August 30, pp. C1, C8.

———. 1996. Biologists Closing on Hidden Lair of Giant Squid. *New York Times* February 13, pp. C1, C9.

———. 1997. *The Universe Below.* Simon & Schuster.

Broch, H. 1954. Blekksprut. *Fauna* 7(4):145–54.

Brooke, A. de C. 1823. *Travels Through Sweden, Norway and Finnmark in the Summer of 1820.* London.

Bruun, A. 1945. Cephalopoda. *Zoology of Iceland* 64(4):1–15.

Buchanan, J. Y. 1896. The Sperm Whale and Its Food. *Nature* 1367(53):223–25.

Buckland, F. T. 1867. *Curiosities of Natural History, Second Series.* Richard Bentley. London.

————. 1883 (published posthumously). *Log-Book of a Fisherman and Zoologist.* Chapman & Hall.

Budelmann, B.-U. 1980. Equilibrium and Orientation in Cephalopods. *Oceanus* 23(3):34–43.

————. 1990. The Statocysts of Squid. In D. L. Gilbert, W. J. Adelman, and J. M. Arnold, eds., *Squid as Experimental Animals,* pp. 421–42. Plenum.

————. 1992. Hearing in Nonarthropod Invertebrates. In D. B. Webster, R. R. Fay, and A. N. Popper, eds., *The Evolutionary Biology of Hearing,* pp. 141–55. Springer-Verlag.

Budelmann, B.-U., and H. Bleckmann. 1988. A Lateral Line Analogue in Cephalopods: Water Waves Generate Microphonic Potentials in the Epidermal Head Lines of *Sepia* and *Lollinguncula. Jour. Comp. Physiol.* 164:1–5.

Budelmann, B.-U., U. Riese, and H. Bleckmann. 1991. Stucture, Function, Biological Significance of the Cuttlefish "Lateral Lines." In E. Boucaud-Comouu, ed., *The Cuttlefish,* pp. 201–09. Centre de Publications de l'Université de Caen.

Bullen, F. T. 1898. *The Cruise of the "Cachalot": Round the World After Sperm Whales.* Appleton.

————. 1902. *Deep-Sea Plunderings.* Appleton.

————. 1904. *Denizens of the Deep.* Revell.

Bulletin of Marine Science. 1991. (Gilbert L. Voss Memorial Issue.) 49(1-2):1–670.

Burkenroad, M. D. 1943. A Possible Function of Bioluminescence. *Jour. Mar. Res.* 5:161–64.

Burton, M. 1957. *Animal Legends.* Coward-McCann.

Buschbaum, R., and L. J. Milne. 1960. *The Lower Animals: Living Invertebrates of the World.* Doubleday.

Buschbaum, R., M. Buschbaum, J. Pearse, and V. Pearse. 1987. *Animals Without Backbones.* University of Chicago Press.

Cadenat, J. 1935. Note sur le première capture dans le Golfe de Goscogne du céphalopode géant *(Architeuthis nawaji). C. r. Ass. fr. Avanc. Sci.* 59:513.

———. 1936. Note sur une céphalopode géant (*Architeuthis harveyi* Verrill) capturé dans le Golfe de Gascogne. *Bull. Mus. Natn. Hist. Nat. Paris* 8:277–85.

Caldwell, D. K., M. C. Caldwell, and D. W. Rice. 1966. Behavior of the Sperm Whale. In K. S. Norris, ed., *Whales, Dolphins and Porpoises*, pp. 677–717. University of California Press.

Canfield, C. 1991. "Giant Squid from Maine a Rare Find." *Portland Press Herald* March 12, pp. 1, 8.

Cheever, H. T. 1859. *The Whale and His Captors*. Nelson.

Chun, C. 1900. *Aus den Tiefen des Weltmeeres*. Gustav Fisher.

———. 1903. Ueber Leuchtorgane und Augen von Tiefsee-Cephalopoden. *Verh. dtsch. zool. Ges.* 13:67–91.

———. 1914. Cephalopoda from the "Michael Sars" North Atlantic Deep-Sea Expedition. *Rep. Sars N. Atl. Deep Sea Exped.* 3:1–28.

Church, R. 1971. *Deepstar* Explores the Ocean Floor. *National Geographic* 139(1):110–129.

Clarke, A., M. R. Clarke, L. J. Holmes, and T. D. Waters. 1985. Calorific Values and Elemental Analysis of Eleven Species of Oceanic Squids (Mollusca: Cephalopoda). *Jour. Mar. Biol. Assoc. U.K.* 65:983–86.

Clarke, A. C. 1953. *Childhood's End*. Ballantine.

———. 1957a. Big Game Hunt. In *Tales from the "White Hart."* Ballantine.

———. 1957b. *The Deep Range*. Harcourt Brace Jovanovich.

———. 1962. The Shining Ones. In *More Than One Universe*. Bantam.

———. 1966. *The Challenge of the Sea*. Dell.

———. 1990. *The Ghost from the Grand Banks*. Bantam.

———. 1992. Squid! A Noble Creature Defended. *Omni* 14(4):70–72.

Clarke, M. R. 1962a. The Identification of Cephalopod "Beaks" and the Relationship Between Beak Size and Total Body Weight. *Bull. Brit. Mus. Nat. Hist. Zool.* 8(10):419–80.

———. 1962b. Stomach Contents of a Sperm Whale Caught off Madeira in 1959. *Norsk Hvalfangst-tidende* 51(5):173–91.

———. 1965. Large Light Organs on the Dorsal Surface of the Squids *Ommastrephes pteropus, Symplectoteuthis oualaniensis,* and *Dosidicus gigas. Proc. Malacol. Soc. London* 36:319–21.

———. 1966. A Review of the Systematics and Ecology of Oceanic Squids. *Adv. Mar. Biol.* 4:91–300.

———. 1967. A Deep-Sea Squid, *Taningia danae* Joubin, 1931. *Symp. Zool. Soc. London* 19:127–43.

———. 1970. The Function of the Spermaceti Organ of the Sperm Whale. *Nature* 228:873–74.

———. 1972. New Techniques for the Study of Sperm Whale Migration. *Nature* 238:405–06.

———. 1976a. Cephalopod Remains from Sperm Whales Caught off Iceland. *Jour. Mar. Biol. Assoc. U.K.* 56:733–49.

———. 1976b. Observation on Sperm Whale Diving. *Jour. Mar. Biol. Assoc. U.K.* 56:809–10.

———. 1977. Beaks, Nets and Numbers. *Sym. Zool. Soc. London* 38:89–126.

———. 1978a. The Cephalopod Statolith—an Introduction to Its Form. *Jour. Mar. Biol. Assoc. U.K.* 58:701–12.

———. 1978b. Structure and Proportions of the Spermaceti Organ in the Sperm Whale. *Jour. Mar. Biol. Assoc. U.K.* 58:1–17.

———. 1979. The Head of the Sperm Whale. *Scientific American* 240(1):128–41.

———. 1980. Cephalopoda in the Diet of Sperm Whales of the Southern Hemisphere and Their Bearing on Sperm Whale Biology. *Discovery Reports* 37:1–324.

———. 1983. Cephalopod Biomass—Estimation from Predation. *Mem. Nat. Mus. Victoria* 44:95–107.

———. 1985. Cephalopods in the Diet of Cetaceans and Seals. *Rapp. Comm. int. Mer Medit.* 29(8):211–19.

———. 1986. *A Handbook for the Identification of Cephalopod Beaks.* Clarendon Press.

———. 1988. Squids. *Biologist* 35(2):69–75.

Clarke, M. R., and C. C. Lu. 1975. Vertical Distribution of Cephalopods at 18°N 25°W in the North Atlantic. *Jour. Mar. Biol. Assoc. U.K.* 55:165–82.

Clarke, M. R., and N. MacLeod. 1976. Cephalopod Remains from Sperm Whales Caught off Iceland. *Jour. Mar. Biol. Assoc. U.K.* 56:733–49.

Clarke, M. R., and N. Merrett. 1972. The Significance of Squid, Whale and Other Remains from the Stomachs of Bottom-Living Deep-Sea Fish. *Jour. Mar. Biol. Assoc. U.K.* 52:599–603.

Clarke, M. R., and P. A. Prince. 1981. Cephalopod Remains in Regurgitations of Black-Browed and Grey-Headed Albatrosses at South Georgia. *Br. Antarct. Surv. Bull.* 54:1–7.

Clarke, M. R., E. J. Denton, and J. B. Gilpin-Brown. 1979. On the Use of Ammonium for Buoyancy in Squids. *Jour. Mar. Biol. Assoc. U.K.* 59:259–76.

Clarke, M. R., J. P. Croxall, and P. A. Prince. 1981. Cephalopod Remains in Regurgitations of the Wandering Albatross *Diomeda exulans* L. at South Georgia. *Br. Antarct. Surv. Bull.* 54:9–21.

Clarke, M. R., N. MacLeod, and O. Paliza. 1976. Cephalopod Remains from the Stomachs of Sperm Whales Caught off Peru and Chile. *Jour. Mar. Biol. Assoc. U.K.* 180:477–493.

Clarke, M. R., N. MacLeod, H. P. Castello, and M. C. Pinedo. 1980. Cephalopod Remains from the Stomach of a Sperm Whale Stranded at Rio Grande do Sul in Brazil. *Mar. Biol.* 59:235–39.

Clarke, R. 1955. A Giant Squid Swallowed by a Sperm Whale. *Norsk Hvalfangst-tidende* 44(10):589–93.

———. 1956. Sperm Whales of the Azores. *Discovery Reports* 28:237–98.

Clarke, W. J. 1933. Giant Squid (New to Science) at Scarborough. *Naturalist* 918:157–58.

Clarke, W. J., and Robson, G. C. 1929. Notes on the Stranding of Giant Squids on the N.E. Coast of England. *Proc. Malacol. Soc.* 18:154–58.

Coe, W. R. 1932. Addison Emery Verrill, 1839–1926. *Biographical Memoirs, National Academy of Sciences* 14:19–66.

Cole, K. S., and D. L. Gilbert. 1970. Jet Propulsion of the Squid. *Biol. Bull.* 138:245–46.

Collett, R. 1912. Odontoceti. In *Norges Pattedyr.* Kristiania.

Conniff, R. 1996a. "The Big Calamari." In *Spineless Wonders,* pp. 58–73. Henry Holt.

————. 1996b. Clyde Roper Can't Wait to Be Attacked by the Giant Squid. *Smithsonian* 27(2):126–37.

Court, W. G. 1980. Japan's Squid Fishing Industry. *Mar. Fish. Rev.* 42(7-8):1–9.

Cousteau, J.-Y. 1954. To the Depths of the Sea by Bathyscaphe. *National Geographic* 106(1):67–79.

Cousteau, J.-Y., and J. Dugan. 1963. *The Living Sea.* Harper & Bros.

Cousteau, J.-Y., and P. Diolé. 1972. *The Whale: Mighty Monarch of the Sea.* Doubleday.

————. 1973a. Last Dance on the Mating Ground. *Natural History* 82(4):45–48.

————. 1973b. *Octopus and Squid: The Soft Intelligence.* Doubleday.

Crichton, M. 1987. *Sphere.* Knopf.

Croker, R. S. 1934. Giant Squid Taken at Laguna Beach. *Calif. Fish and Game* 20(3):297.

————. 1937. Further Notes on the Jumbo Squid, *Dosidicus gigas. Calif. Fish and Game* 23(3):246–47.

Crosse, H., and P. Fischer. 1862. Nouveaux documents sur les céphalopodes gigantesques. *Jour. Conch. Paris* 10:124–40.

Dahlgren, U. 1916. The Production of Light by Animals. Light Production in Cephalopods. *Jour. Franklin Inst.* 81:525–56.

Dall, W. H. 1873. Aleutian Cephalopods. *American Naturalist* 7:484–85.

————. 1885. The Arms of the Octopus, or Devil Fish. *Science* 6:432.

Darling, J. D., C. Nicklin, K. S. Norris, H. Whitehead, and B. Würsig. 1995. *Whales, Dolphins and Porpoises.* National Geographic Society.

Dell, R. K. 1952. The Recent Cephalopoda of New Zealand. *Dominion Mus. Bull.* 16:1–157.

————. 1970. A Specimen of the Giant Squid *Architeuthis* from New Zealand. *Rec. Dominion Mus.* 7(4):25–36.

Delworth, T., S. Manabe, and R. J. Stouffer. 1993. Interdecal Variations of the Thermohaline Circulation of a Coupled Ocean-Atmosphere Model. *Jour. of Climate* 6(11):1993–2011.

De Montfort, P. D. 1802. *Histoire naturelle générale et particulière des mollusques.* From Sonnini's expanded edition of Buffon's *Natural History.* Paris.

Denton, E. J. 1960a. The Buoyancy of Marine Animals. *Scientific American* 203:118–28.

———. 1960b. The "Design" of Fish and Cephalopod Eyes in Relation to Their Environment. *Symp. Zool. Soc. London* 3:53–55.

———. 1974. On Buoyancy and the Lives of Modern and Fossil Cephalopods. *Proc. Roy. Soc. London* 185(1080):273–99.

Denton, E. J., and F. J. Warren. 1968. Eyes of the Histioteuthidac. *Nature* 219:400–01.

Denton, E. J., J. B. Gilpin-Brown, and T. I. Shaw. 1969. A Buoyancy Mechanism Found in Cranchid Squid. *Proc. Roy. Soc. London* 174:271–79.

Doino, J. A., and M. J. McFall-Ngai. 1995. A Transient Exposure to Symbiosis-Competent Bacteria Induces Light Organ Morphogenesis in the Host Squid. *Biol. Bull.* 189:347–55.

Donovan, D. T. 1977. Evolution of the Dibranchiate Cephalopoda. In M. Nixon and J. B. Messenger, eds., *The Biology of Cephalopods*, pp. 15–48. Academic Press.

Douglas, C. 1770. An Account of the Result of Some Attempts to Ascertain the Temperature of the Sea in Great Depths, near the Coasts of Lapland and Norway; and Also Some Anecdotes. *Phil. Trans. Roy. Soc. London* 60:39–43.

Dozier, T. A. 1976. *Dangerous Sea Creatures.* Time-Life Books. Alexandria, Va.

Drevar, G. 1875. The Great Sea Serpent. *Illustrated London News* 10 November: 448–49.

———. 1877. The Great Sea Serpent. *Illustrated London News* 13 January: 78.

Duncan, D. 1941. Fighting Giants of the Humboldt. *National Geographic* 79(3):373–400.

Duncan, N. 1940. The Adventure of the Giant Squid of Chain Tickle. In E. Johnson and C. E. Scott, eds., *Anthology of Children's Literature*, pp. 621–23. Houghton Mifflin. Boston.

Earle, S. A. 1991. Sharks, Squids, and Horseshoe Crabs—the Significance of Marine Biodiversity. *Bioscience* 41(7):506–10.

Egede, H. 1745. *A Description of Greenland.* London.

Ellis, R. 1980. *The Book of Whales.* Knopf.

———. 1982. *Dolphins and Porpoises.* Knopf.

———. 1994a. *Architeuthis*—the Giant Squid: A True Explanation for Some Sea Serpents. *Log of Mystic Seaport* 46(2):34–40.

———. 1995. *Monsters of the Sea*. Knopf.

———. 1996. *Deep Atlantic*. Knopf.

Emboden, W. 1974. Our Sea Serpent. *Terra. Q. Mag. Nat. Hist. Mus. Los Angeles Co.* 12(3):12–18.

Emerton, J. H. 1883. Model of the Giant Octopus of the West Coast of North America. *Science* 1(13):352–55.

Ferguson, G. P., and J. B. Messenger. 1991. A Countershading Reflex in Cephalopods. *Proc. Roy. Soc. London* (243):63–67.

Filippova, J. A. 1972. New Data on the Squids from the Scotia Sea (Antarctic). *Malacologia* 11(2):391–406.

Fiscus, C. H., and D. W. Rice. 1974. Giant Squids, *Architeuthis* sp., from Stomachs of Sperm Whales Captured off California. *Calif. Fish and Game* 60(2):91–93.

Fiscus, C. H., D. W. Rice, and A. A. Wolman. 1989. Cephalopods from the Stomachs of Sperm Whales Taken off California. *NOAA Technical Report NMFS* 83:1–12.

Fisher, A. 1995. He Seeks the Giant Squid. *Popular Science* 246(5):28–31.

———. 1997. The Hunt for the Giant Squid: A High-Tech Search for the Sea's Most Elusive Creature. *Popular Science* 248(3):74–77.

Fjelstrup, A. 1887. Hudens bygning hos *Globiocephalas melas*. *Vidensk. Meddel. Dansk naturh. Foren.* 39:227–35.

Fleischer, R. 1993. *Just Tell Me When to Cry: A Memoir*. Carroll & Graf.

Fleisher, K. J., and J. F. Case. 1995. Cephalopod Predation Facilitated by Dinoflagellate Bioluminescence. *Biol. Bull.* 189:263–71.

Fleming, I. 1958. *Doctor No*. Cape.

Foyle, T. P., and R. K. O'Dor. 1987. Predatory Strategies of Squid *(Illex illecebrosus)* Attacking Small and Large Fish. *Mar. Behav. Physiol.* 13:155–68.

Frazier, J., and H. Hathorne. 1984. *20,000 Leagues Under the Sea: The Filming of Jules Verne's Classic Science Fiction Novel*. *Cinéfantastique* 14(3):32–53.

Frost, N. 1934. Notes on a Giant Squid (*Architeuthis* sp.) Captured at Dildo, Newfoundland, in December 1933. *Rep. Nfld. Fish. Comm.* 2(2):100–14.

———. 1936. A Further Species of Giant Squid (*Architeuthis* sp.) from Newfoundland Waters. *Rep. Nfld. Fish. Comm.* 2(5):89–95.

Furtado, A. 1887. Sur une nouvelle espèce de céphalopode appertenat au genre *Ommastrephes. Mem. R. Acad. Lisboa.* 3–18.

Gage, J. D., and P. A. Tyler. 1991. *Deep-Sea Biology: A Natural History of Organisms at the Deep-Sea Floor.* Cambridge University Press.

Gaskin, D. E. 1967. Luminescense in the Squid *Moroteuthis* (Probably *ingens* Smith) and a Possible Feeding Mechanism in the Sperm Whales *Physeter catadon. Tuatara* 15:86–88.

Gaskin, D. E., and M. W. Cawthorn. 1967. Diet and Feeding Habits of the Sperm Whale (*Physeter catodon* L.) in the Cook Strait Region of New Zealand. *N.Z. Jour. Mar. Freshwater Res.* 1(2):159–79.

Gauldie, R. W., I. F. West, and E. C. Förch. 1994. Statocyst, Statolith, and Age Estimation of the Giant Squid, *Architeuthis kirki. Veliger* 37(1):93–109.

Gervais, P. 1875. Remarques au sujet des grands céphalopodes décabranchidés. *Jour. Zoologie* 4:88–95.

Gesner, C. 1551–58. *Historia animalium.* Zurich.

———. 1551–58. *De piscium et aquatilium animatum natura.* Zurich.

Gibson, J. 1887. *Monsters of the Sea, Legendary and Authentic.* Thomas Nelson & Sons.

Gilpin-Brown, J. B. 1977. The Squid and Its Giant Nerve Fibre. In M. Nixon and J. B. Messenger, eds., *The Biology of Cephalopods,* pp. 233–41. Academic Press.

Girard, A. A. 1892. Les Céphalopodes des Îles Açores et de l'Île de Madere. *Jour. Sci. Math. Phys. Nat.* Series 2, No. 6:210–20.

Gosline, J. M., and M. E. DeMont. 1984. Jet-Propelled Swimming in Squids. *Scientific American* 252(1):96–103.

Gosse, P. H. 1861. *The Romance of Natural History.* James Nisbet.

Gould, C. 1886. *Mythical Monsters.* London. (Crescent Books ed., New York, 1989.)

Gould, R. T. 1930. *The Case for the Sea Serpent.* Phillip Allan.

Greenwell, J. R. 1983. "Sea Serpents" Seen off California Coast. *ISC Newsletter* 2(4):9–10.

———. 1985. Giant Octopus Blamed for Deep-Sea Fishing Disruptions. *ISC Newsletter* 4(3):1–6.

Grieg, J. A. 1933a. The Cephalopod Fauna of Svalbard. *Tromsö Mus. Arsh.* 53:1–19.

———. 1933b. Cephalopods from the West Coast of Norway. *Bergens. Mus. Aarb.* 1(4):1–25.

Grimpe, G. 1925. Zur Kenntnis der Cephalopodenfauna der Nordsee. *Wiss. Meeresunters.* 16:1–124.

Grønningsaeter, A. 1946. Sjørmen-blekksprutten. *Naturen* 70:379–80.

Haimovici, M., and J. A. A. Perez. 1991. Coastal Cephalopod Fauna of Southern Brazil. *Bull. Mar. Sci.* 49(1-2):221–30.

Hall, A. 1976. *Monsters and Mythic Beasts.* Doubleday.

Hamabe, M., C. Hamuro and M. Ogura. 1982. *Squid Jigging from Small Boats.* FAO Fisheries Technology Service. Fishing New Books.

Hamilton, J. E. 1914. Belmullet Whaling Station. Report of the Commission. *Rep. Br. Assoc. Adv. Sci.* 125–61, 137–8.

Hanlon, R. T. 1987. Traditional Squid Fishing in the Azores. *Sea Frontiers* 33(1):34–41.

Hanlon, R. T., and B.-U. Budelmann. 1987. Why Cephalopods Are Probably Not "Deaf." *American Naturalist* 129(2):312–17.

Hanlon, R. T., and J. B. Messenger. 1996. *Cephalopod Behaviour.* Cambridge University Press.

Hardy, A. C. 1956. *The Open Sea: Its Natural History. Part I: The World of Plankton.* Houghton Mifflin.

———. 1967. *Great Waters.* Harper & Row.

Hart, H. M. 1882. *The World of the Sea.* Cassell, Petter, Galpin & Co.

Harting, P. 1860. Description des quelques fragments de deux céphalopodes gigantesques. *Naturk. Verh. Koninkl. Akad. Wet.* 9:1–16.

Harvey, E. N. 1952. *Bioluminscence.* Academic Press.

Harvey, M. 1874. Gigantic Cuttlefishes in Newfoundland. *Ann. Mag. Nat. Hist.* 13:67–70.

———. 1879. Untitled Article in *Boston Traveller,* January 30.

———. 1899. How I Discovered the Great Devil-Fish. *Wide World Magazine* 2:732–40.

Hauser, H. 1987. California Diver Bags Giant Squid. *Skin Diver* 36(9):110–11.

Heath, H. 1917. Devilfish and Squid. *Calif. Fish and Game* 3(3):1–6.

Hedd, A., R. Gales, N. Brothers, and G. Robertson. 1997. Diving Behaviour of the Shy Albatross *Diomeda cauta* in Tasmania: Initial Findings and Dive Recorder Assessment. *Ibis.* In press.

Heezen, B. C. 1957. Whales Entangled in Deep-Sea Cables. *Deep-Sea Res.* 4:105–15.

Heezen, B. C., and C. D. Hollister. 1971. *The Face of the Deep.* Oxford University Press.

Heilner, V. C. 1953. *Salt Water Fishing.* Knopf.

Hellman, G. 1968. *Bankers, Bones and Beetles: The First Century of the American Museum of Natural History.* Natural History Press.

Helm, T. 1962. *Monsters of the Deep.* Dodd, Mead.

Heppell, D. 1977. Giant Squid Stranded at North Berwick. *Porcupine Newsletter* 1:63.

———. 1978. Giant Squid Stranded at North Berwick. *Conchologist's Newsletter* 65:89.

Heppell, D., and S. M. Smith. 1983. Recent Cephalopoda in the Collections of the Royal Scottish Museum, Edinburgh. *Royal Scottish Museum Information Series (Natural History)* 10:1–81.

Herdman, W. A. 1956. The Prince of Monaco and the Oceanographic Museum. *Bull. Trimestriel Musée Océanographique de Monaco* 38:1–12.

Herring, P. J. 1977. Luminescence in Cephalopods and Fish. In M. Nixon and J. B. Messenger, eds., *The Biology of Cephalopods,* pp. 127–60. Academic Press.

Herring, P. J., and M. R. Clarke, eds. 1971. *Deep Oceans.* Praeger.

Herring, P. J., P. N. Dilly, and C. Cope. 1992. Different Types of Photophores in the

Oceanic Squids *Octopoteuthis* and *Taningia* (Cephalopoda: Octopoteuthidae). *Jour. Zool. London* 227:479–91.

Heuvelmans, B. 1958. *Dans le sillage des monstres marins: Le kraken et le poulpe colossal.* Librairie Plon.

——. 1965. *In the Wake of the Sea-Serpents.* Hill & Wang.

——. 1990. The Metamorphosis of Unknown Animals into Fabulous Beasts and of Fabulous Beasts into Known Animals. *Cryptozoology* 9:1–12.

——. 1995. *On the Track of Unknown Animals.* Kegan Paul.

——. 1997. *The Kraken and the Colossal Octopus.* Kegan Paul.

Hilgendorf, F. 1880. Uber einen riesigen Tintenfisch aus Japan, *Megateuthis martensii* g.n., sp.n. *S.B. Ges. naturf. Fr. Berlin* 65–7.

Hillinger, C. 1974. Biologist Plans Ocean Search for Giant Squid. *Los Angeles Times* May 5, pp. ix, 1–2.

Hochberg, F. G. 1974. Southern California Records of the Giant Squid, *Moroteuthus robusta. Tabulata* 7(4):83–85.

——. 1986. Of Beaks and Whales. *Bull. Santa Barbara Mus. Nat. Hist.* 95:1–2.

——. 1988. The Skin of Cephalopods—a Living Tapestry of Color and Texture. *Bull. Santa Barbara Mus. Nat. Hist.* 122:1.

Hochberg, F. G., and W. G. Fields. 1980. Cephalopoda: The Squids and Octopuses. In R. H. Morris, D. P. Abbot, and C. D. Haderlie, eds., *Intertidal Invertebrates of California,* pp. 429–44. Stanford University Press.

Holder, C. F. 1899. Some Pacific Cephalopods. *Scientific American* 80:253.

Holthe, T. 1975. A New Record of a Giant Squid from Trondheimsfjorden. *Fauna* 28(3):174.

——. 1976. Kjempeblekksprut. *Addresseavisen* 22(3):29.

Homer. N.d. *The Odyssey.* (Translated by Robert Fitzgerald, Doubleday Anchor, New York, 1963.)

Honma, Y. 1990. Revised Records of Large Marine Animals Stranded on the Coast of Niigata and Sado Island in the Sea of Japan. *Spec. Publ. Sado Mar. Biol. Stat. Niigata Univ.* 5:1–39.

Honma, Y., T. Kitami, and R. Mizusawa. 1983. Record of Cephalopoda in the

Waters Adjacent to Niigata and Sado Island in the Japan Sea, Based Partially on the Pelagic Squids Stranded Ashore. *Bull. Biogeogr. Soc. Japan* 38(3):23–29.

Hoskin, F. C. G. 1990. An Organophosphorus Detoxifying Enzyme Unique to Squid. In D. L. Gilbert, W. J. Adelman, and J. M. Arnold, eds., *Squid as Experimental Animals,* pp. 469–80. Plenum.

Hoskin, F. C. G., and A. H. Roush. 1982. Hydrolysis of Nerve Gas by Squid-Type Disopropyl Phosphorofluoridate Hydrolyzing Enzyme on Agarose Resin. *Science* 215:1255–57.

Hoskin, F. C. G., and J. E. Walker. 1995. What the Squid Knows About Nerve Gas. Unpublished ms.

Houot, G. S., and P. H. Willm. 1955. *2,000 Fathoms Down.* Dutton.

Hoyle, W. E. 1886. Report on the Cephalopoda. *Rep. Scient. Results H.M.S. "Challenger" Exped., Zool.* 16:1–245.

———. 1904. Reports on the Cephalopoda from the "Albatross" Expedition. *Bull. Mus. Comp. Zool.* 43(1):1–71.

———. 1908. A Large Squid at Redcar. *Naturalist* 615:132–33.

Hunt, J. C. 1997. *Octopus and Squid.* Monterey Bay Aquarium.

Idyll, C. P. 1971. *Abyss: The Deep Sea and the Creatures That Live in It.* Thomas Y. Crowell.

Imber, M. J. 1976. Comparison of Prey of the Black *Procellaria* Petrels of New Zealand. *N.Z. Jour. Mar. Freshwater Res.* 10(1):199–30.

———. 1978. The Squid Families Cranchiidae and Gonatidae in the New Zealand Area. *N.Z. Jour. Zool.* 5(3):445–84.

———. 1992. Cephalopods Eaten by Wandering Albatrosses (*Diomeda exulans* L.) Breeding at Six Circumpolar Localities. *Jour. Roy. Soc. N.Z.* 22(4):243–63.

Imber, M. J., and R. Russ. 1975. Some Foods of the Wandering Albatross (*Diomeda exulans*). *Notorinis* 22(1):27–36.

Ishiwaka, C., and Y. Wakiya. 1914a. Note on a Gigantic Squid Obtained from the Stomach of a Sperm Whale. *Jour. Coll. Agric. Imp. Univ. Tokyo* 4:435–43.

———. 1914b. On a New Species of *Moroteuthis* from the Bay of Sagami, *M. lonnbergi. Jour. Coll. Agric. Imp. Univ. Tokyo* 4:445–60.

Iverson, I. L., and L. Pinkas. 1971. A Pictorial Guide to Beaks of Certain Eastern Pacific Cephalopods. *Calif. Dept. Fish and Game Fish. Bull.* 152:83–105.

Iverson, R. T. S., P. J. Perkins, and R. D. Dionne. 1963. An Indication of Underwater Sound Production by Squid. *Nature* 199:250–51.

Iwai, E. 1956. Descriptions of Unidentified Species of Dibranchiate Cephalopods I. An Oegopsiden Squid Belonging to the Genus *Architeuthis*. *Sci. Rep. Whales Res. Inst.* 11:139–51.

Jackson, G. D., C. C. Lu, and M. Dunning. 1991. Growth Rings Within the Statolith Microstructure of the Giant Squid *Architeuthis*. *Veliger* 34(4):331–34.

Jameson, W. 1958. *The Wandering Albatross*. Anchor.

Jeffreys, I. G. 1869. *British Conchology*. London.

Jensen, A. S. 1915. On Some Misinterpreted Markings on the Skin of the Caaing Whale. *Vidensk. Meddel. Dansk naturh. Foren. Bd.* 67:1–8.

———. 1916. Addendum to My Paper: On Some Misinterpreted Markings, Etc. *Vidensk. Meddel. Dansk naturh. Foren. Bd.* 67:221–22.

Johnson, R. I. 1989. Molluscan Taxa of Addison Emery Verrill and Katharine Jeannette Bush, Including Those Introduced by Sanderson Smith and Alpheus Hyatt Verrill. *Occasional Papers on Molluscs, Museum of Comparative Zoology* 5(67):1–80.

Jonstonus, J. 1649. *Historia naturalis*. Frankfurt.

Joubin, L. 1902. Observations sur divers céphalopodes sixième note: Sur une nouvelle espèce du genre *Rossia*. *Bull. Soc. Zool. Fr.* 27:138–43.

———. 1931. Notes préliminaires sur les céphalopodes des croisières du "Dana" (1921–1922). *Ann. Inst. Océanogr.* 10(7):167–211.

———. 1933. Notes préliminaires sur les céphalopodes des croisières du "Dana" (1921–1922). *Ann. Inst. Océanogr.* 13(1):1–49.

———. 1937. Les octopodes de la croisière du "Dana," 1921–22. *Dana Report* 11:1–49.

Kawakami, T. 1976. Squids Found in the Stomach of Sperm Whales in the Northwestern Pacific. *Sci. Rep. Whales Res. Inst.* 28:145–51.

———. 1980. A Review of Sperm Whale Food. *Sci. Rep. Whales Res. Inst.* 32:199–218.

Kemp, M. 1989. A Squid for All Seasons. *Discover* 10(6):66–70.

Kenney, J. 1978. Tall Tales and Sea Serpents. *Oceans* 11(4):8–12.

Kent, W. S. 1874a. Note on a Gigantic Cephalopod from Conception Bay, Newfoundland. *Proc. Zool. Soc. London* 1874:178–82.

———. 1874b. A Further Communication upon Certain Gigantic Cephalopods Recently Encountered off the Coast of Newfoundland *Proc. Zool. Soc. London* 1874:489–94.

Kier, W. M., and A. M. Smith. 1990. The Morphology and Mechanics of Octopus Suckers. *Biol. Bull.* 178:126–36.

Kirk, T. W. 1880. On the Occurrence of Giant Cuttlefish on the New Zealand Coast. *Trans. N.Z. Inst.* 12:310–13.

———. 1888. Brief Description of a New Species of Large Decapod *(Architeuthis longimanus)*. *Trans. N.Z. Inst.* 20:34–39.

Kjennerud, J. 1958. Description of a Giant Squid, *Architeuthis,* Stranded on the West Coast of Norway. *Universitet i Bergen Arbok 1958 Naturv. Rekke* 9:3–14.

Knudsen, J. 1957. Some Observations on a Mature Male Specimen of *Architeuthis* from Danish Waters. *Proc. Malacol. Soc. London* 32:189–98.

Koefoed, E. 1950. Bleksprutter. In B. Føyn, G. Ruud, H. Røise, and H. Christensen, eds., *Norges Dyreliv IV* (Animal Life of Norway, vol. 4), pp. 420–27. Cappelens.

Kozak, V. A. 1974. Receptor Zone of the Video-Acoustic System of the Sperm Whale *(Physter catodon* L. 1758). *Fiziologichnyy Zhurnal Akademy Nauk Ukrayns'koy RSR* 120(3):1–6. (National Technical Information Service, 1975.)

Kristensen, T. K., and J. Knudsen. 1983. A Catalogue of the Type Specimens of Cephalopoda (Mollusca) in the Zoological Museum, University of Copenhagen. *Steenstrupia* 9(10):217–27.

Lamy, E. 1935. Louis Joubin (1865–1935). *Jour. de Conchyliologie* 79:204–09.

Lane, F. W. 1963. Brilliance in the Ocean Depths. *Animals* 2(2):30–33.

———. 1974. *The Kingdom of the Octopus.* Sheridan House. New York.

Lange, M. M. 1920. On the Regeneration and Finer Structure of the Arms of the Cephalopods. *Jour. Exp. Biol.* 31:1–40.

LeBlond, P. H., and E. L. Bousfield. 1995. *Cadborosaurus: Survivor from the Deep.* Horsdal & Schubart.

LeBlond, P. H., and J. Sibert. 1973. Observations of Large Unidentified Marine Animals in British Columbia and Adjacent Waters. Unpublished ms. Institute of Oceanography, University of British Columbia.

Lee, H. 1875. *The Octopus.* London.

———. 1884a. *Sea Fables Explained.* London.

———. 1884b. *Sea Monsters Unmasked.* London.

Lewis, T. A. 1990. Squid: The Great Communicator? *National Wildlife* 28(5):14–19.

Ley, W. 1941. Scylla Was a Squid. *Natural History* 48(1):11–13.

———. 1948. *The Lungfish, the Dodo, and the Unicorn.* Viking Press.

———. 1968. *Dawn of Zoology.* Prentice-Hall.

———. 1987. *Exotic Zoology.* Bonanza.

Linnaeus, C. 1758. *Systema naturae.* 10th ed. *Vol. I: Regnum animale.* Holmiae.

Lipinski, M. 1986. Methods for the Validation of Squid Age from Statoliths. *Jour. Mar. Biol. Assoc. U.K.* 66:505–26.

Lipinski, M. R., and S. Jackson. 1989. Surface-Feeding on Cephalopods by Procellariiform Seabirds in the Southern Benguela Region, South Africa. *Jour. Zool.* 218:549–63.

Lipinski, M., and K. Turoboyski. 1983. The Ammonium Content in the Tissues of Selected Species of Squid (Cephalopoda: Teuthoidea). *Jour. Exp. Mar. Biol. Ecol.* 69:145–50.

Lockyer, C. 1997. Diving Behavior of the Sperm Whale in Relation to Feeding. In T. G. Jacques and R. H. Lambertsen, eds., *Sperm Whale Deaths in the North Sea: Science and Management,* pp. 47–52. *Bull. Inst. Royal Sci. Nat. Belg.* 67: Supplement.

Lonnberg, E. 1891. Ofversigt ofver Sveriges Cephalopoden. *Bih. k. Svensk. Vct. Ak. Handl. Bd.* 17:1–42.

Lu, C. C. 1977. A New Species of Squid, *Chiroteuthis acanthoderma* from the Southwest Pacific (Cephalopoda: Chiroteuthidae). *Steenstrupia* 4:179–88.

———. 1986. Smallest of the Largest: First Record of Giant Squid Larval Specimen. *Australian Shell News* 53:9.

Lu, C. C., and M. R. Clarke. 1975a. Vertical Distribution of Cephalopods at 40°N, 53°N and 60° at 20°W in the North Atlantic. *Jour. Mar. Biol. Assoc. U.K.* 55:143–63.

———. 1975b. Vertical Distribution of Cephalopods at 11°N, 20°W in the North Atlantic. *Jour. Mar. Biol. Assoc. U.K.* 55:369–89.

Lu, C. C., and C. F. E. Roper. 1979. Cephalopods from Dumpsite 106 (Western Atlantic): Vertical Distribution and Seasonal Abundance. *Smithsonian Contributions to Zoology* 288:1–36.

MacGintie, G. E., and N. MacGintie. 1949. *Natural History of Marine Animals.* McGraw-Hill.

MacKintosh, N. A. 1956. 2,000 Feet Down: Deep-Sea Photographs Teach Us More About the Little-Known Squid. *New Scientist* 1:60.

MacLeish, W. H. 1980. Mysterious Creatures. In J. J. Thorndike, ed., *Mysteries of the Deep,* pp. 242–91. American Heritage.

Macrae, J., and Twopeny, D. 1873. Appearance of an Animal, Believed to Be That Which Is Called the Norwegian Sea Serpent, on the Western Coast of Scotland, in August, 1872. *Zoologist* 1(8):988–93.

Magnus, O. 1555. *Historia de gentibus septentrionalibus.* Antwerp.

Mangold, K. 1983. Food, Feeding, and Growth in Cephalopods. *Mem. Nat. Mus. Victoria* 44:81–93.

Marshall, N. B. 1979. *Deep-Sea Biology.* Garland.

Martin, R. 1977. The Giant Nerve Fibre System of Cephalopods. Recent Structural Findings. In M. Nixon and J. B. Messenger, eds., *The Biology of Cephalopods,* pp. 261–75. Academic Press.

Massy, A. L. 1928. The Cephalopoda of the Irish Coast. *Proc. R. Irish Acad.* 38:25–37.

Matthews, L. H. 1938. The Sperm Whale, *Physeter catadon. Discovery Reports* 27:93–168.

Maturana, H. R., and S. Sperling. 1963. Unidirectional Response to Angular Acceleration Recorded from the Middle Cristal Nerve in the Statocyst of *Octopus vulgaris. Nature* (London) 197:815–16.

Mauro, A. 1977. Extra-Ocular Receptors in Cephalopods. In M. Nixon and J. B. Messenger, eds., *The Biology of Cephalopods,* pp. 287–308. Academic Press.

McCann, C. 1974. Body Scarring on Cetaceans—Odontocetes. *Sci. Rep. Whales Res. Inst.* 26:145–55.

McCapra, F. 1990. The Chemistry of Bioluminescence: Origins and Mechanisms. In P. J. Herring, A. K. Campbell, M. Whitfield, and L. Maddock, eds., *Light and Life in the Sea,* pp. 221–27. Cambridge University Press.

McDowall, C. A. 1998. Giant Squid? *Marine Observer* 68 (339):39.

McEwan, G. J. 1978. *Sea Serpents, Sailors, and Skeptics.* Routledge & Kegan Paul.

McSweeny, E. S. 1970. Description of the Juvenile Form of *Mesonychoteuthis hamiltoni* Robson. *Malacologia* 10:323–32.

———. 1978. Systematics and Morphology of the Antarctic Cranchid Squid *Galiteuthis glacialis* (Chun). *Antarc. Res. Ser.* 27:1–39.

Meade-Waldo, E. G. B., and M. J. Nicoll. 1906. Description of an Unknown Animal Seen at Sea off the Coast of Brazil. *Proc. Zool. Soc. London* 2(98):719.

Meek, A., and T. R. Goddard. 1926. On Two Specimens of Giant Squid Stranded on the Northumberland Coast. *Trans. Nat. Hist. Soc. Newcastle* 6:229–37.

Meinertzhagen, I. A. 1990. Development of the Squid's Visual System. In D. L. Gilbert, W. J. Adelman, and J. M. Arnold, eds., *Squid as Experimental Animals,* pp. 399–419. Plenum.

Melville, H. 1851. *Moby-Dick.* New York. (Norton Critical Ed., H. Hayford and H. Parker, eds. W. W. Norton.)

Mercer, R. W., and M. Bucy. 1983. Experimental Squid Jigging off the Washington Coast. *Mar. Fish. Rev.* 45(7-8-9):56–62.

Messenger, J. B. 1977. Prey-Capture and Learning in the Cuttlefish, *Sepia.* In M. Nixon and J. B. Messenger, eds., *The Biology of Cephalopods,* pp. 347–76. Academic Press.

Millam, G. 1984. Tripping the Light Fantastic: The Sensitive World of Bioluminescence. *Oceans* 17(4):3–8.

Miller, J. A. 1983. Super Squid Lies in State. *Science News* 123(7):110.

Miller, W. J. 1976. *The Annotated Jules Verne: Twenty Thousand Leagues Under the Sea.* Crowell.

Milne, L. J. 1947. Squid. *Atlantic Monthly* 180(2):104–06.

Miner, R. W. 1935. Marauders of the Sea. *National Geographic* 68(2):185–207.

Mitsukuri, K., and S. Ikeda. 1895. Note on a Giant Cephalopod. *Zool. Mag. Tokyo* 7:39–50.

Mizue, K. 1951. Food of Whales (in the Adjacent Waters of Japan). *Sci. Rep. Whales Res. Inst.* 5:81–90.

Moiseev, S. I. 1991. Observation of the Vertical Distribution and Behavior of Nektonic Squids Using Manned Submersibles. *Bull. Mar. Sci.* 49(1-2):446–56.

More, A. G. 1875a. Gigantic Squid on the West Coast of Ireland. *Ann. Mag. Nat. Hist.* 4(16):123–24.

———. 1875b. Notice of a Gigantic Cephalopod *(Dinoteuthis proboscideus)* Which Was Stranded at Dingle, in Kerry, Two Hundred Years Ago. *Zoologist* 2(10):4526–32.

———. 1875c. Some Account of the Gigantic Squid *(Architeuthis dux)* Lately Captured off Boffin Island, Connemara. *Zoologist* 2(11):4569–71.

Morrison, P. 1996. Giant Against Giant in the Dark. *Scientific American* 275(5):124, 126.

Morton, J. E., and C. M. Yonge. 1964. Classification and Structure of Molluscs. In K. M. Wilbur and C. M. Yonge, eds., *Physiology of Mollusca,* vol. 1, pp. 1–58. Academic Press.

Moynihan, M. 1975. Conservatism of Displays and Comparable Stereotyped Patterns Among Cephalopods. In G. Baerends, C. Beer, and A. Manning, eds., *Function and Evolution in Behaviour: Essays in Honour of Professor Niko Tinbergen, F.R.S.,* pp. 276–91. Oxford University Press.

———. 1983. Notes on the Behavior of *Euprymna scolopes* (Cephalopoda: Sepiolidae). *Behaviour* 85:25–41.

———. 1985a. *Communication and Noncommunication by Cephalopds.* Indiana University Press.

———. 1985b. Why Are Cephalopods Deaf? *American Naturalist* 125(3):465–69.

Moynihan, M., and A. F. Rodaniche. 1977. Communication, Crypsis, and Mimicry Among Cephalopods. In T. Sebeok, ed., *How Animals Communicate,* pp. 293–302. Indiana University Press.

———. 1982. *The Behavior and Natural History of the Caribbean Reef Squid Sepioteuthis sepioidea, with a Consideration of Social, Signal, and Defensive Patterns for Difficult and Dangerous Environments.* Paul Parey.

Muntz, W. R. A. 1977. Pupillary Response of Cephalopods. In M. Nixon and J. B. Messenger, eds., *The Biology of Cephalopods*, pp. 277–85. Academic Press.

———. 1995. Giant Octopuses and Squid from Pliny to the Rev. Moses Harvey. *Archives of Natural History* 22(1):1–28.

Murie, J. 1865. On Deformity of the Lower Jaw in the Cachalot. *Proc. Zool. Soc. London* 1865:390–96.

Murray, A. 1874a. Capture of a Giant Squid at Newfoundland. *American Naturalist* 8(2):120–23.

———. 1874b. Notice of a Gigantic Squid. *Proc. Boston Soc. Nat. Hist.* 16:161–63.

Murray, J., and J. Hjort. 1912. *The Depths of the Ocean*. Macmillan & Co. London.

Muus, B. J. 1959. Skallus, Søtaender, Blaeksprutter. *Danmarks Fauna* 65:1–239.

———. 1962. Cephalopoda. The Godthaab Expedition 1928. *Meddr. Grönland* 81:4–21.

Myklebust, B. 1946. Et nytt funn av kjempeblekksprut i Romsdal. *Naturen* 70:377–79.

Nasu, K. 1958. Deformed Lower Jaw of the Whale. *Sci. Rep. Whales Res. Inst.* 13:211–12.

Nemoto, T. 1957. Foods of Baleen Whales in the Northern Pacific. *Sci. Rep. Whales Res. Inst.* 12:33–89.

Nemoto, T., and K. Nasu. 1963. Stones and Other Aliens in the Stomachs of Sperm Whales in the Bering Sea. *Sci. Rep. Whales Res. Inst.* 17:83–91.

Nesis, K. N. 1970. The Biology of the Giant Squid of Peru and Chile, *Dosidicus gigas*. *Okeanologiia* 10:140–52. (In Russian; English summary.)

———. 1974. Giant Squids. *Priroda* 6:55–60.

———. 1975. A Revision of the Squid Genera *Corynomma, Megalocranchia, Sandalops,* and *Liguriella* (Oegopsida, Cranchiidae). *Tr. Inst. Okeanol. Acad. Nauk SSSR* 96:6–22. (Translation by Canada Institute for Scientific and Technical Information.)

———. 1977. Vertical Distribution of Pelagic Cephalopods. *Jour. General Biol. (Zhurnal Obshchei Biologii.)* 38(4):547–57. (1979 translation by National Museums Canada Translation Bureau.)

———. 1982. *Cephalopods of the World.* T. F. H. Publications, Neptune City, N.J. (Translated from the Russian by B. S. Levitov.)

———. 1983. *Dosidicus gigas.* In P. R. Boyle, ed., *Cepahalopod Life Cycles, Vol. 1: Species Accounts,* pp. 215–31. Academic Press.

———. 1985. A Giant Squid in the Sea of Okhotsk. *Priroda* 10:112–13. (Translated from the Russian by Yuri Nektorenko.)

———. 1991. Cephalopods of the Benguela Upwelling off Namibia. *Bull. Mar. Sci.* 49(1-2):199–215.

———. 1992. A Window to the Ocean: Giants and Dwarves of the Soft Intelligence. *Rybak Primorye.* (Maritime Fisherman.) 26 June 325(564):6. (Vladivostok newspaper translated by Yuri Nekrutenko.)

———. 1993. Okhotsk Sea—Home of Giant Cephalopods. (Abstract.) In T. Okutani, R. K. O'Dor, and T. Kubodera, eds., *Recent Advances in Cephalopod Fisheries Biology,* pp. 745–46. Tokai University Press.

Nesis, K. N., A. M. Amelekhina, A. R. Boltachev, and G. A. Shevtsov. 1985. Records of Giant Squids of the Genus *Architeuthis* in the North Pacific and South Atlantic. *Zoologicheskiy Zhurnal* 64(4):518–28.

Nichols, A. R. 1905. On Some Irish Specimens of a Large Squid, *Sthenoteuthis pteropus* (Steenstrup). *Irish Naturalist* 14:54–57.

Nicol, J. A. C. 1958. Observations on Luminescence in Pelagic Animals. *Jour. Mar. Biol. Assoc. U.K.* 37:705–22.

———. 1961. Luminescence in Marine Organisms. *Smithsonian Rep.* 1960:447–56.

———. 1964a. Luminous Creatures of the Sea. *Sea Frontiers* 10(3):143–54.

———. 1964b. Special Effectors: Luminous Organs, Chromatophores, Pigments, and Poison Glands. In K. M. Wilbur and C. M. Yonge, eds., *Physiology of Mollusca,* vol. 1, pp. 353–81. Academic Press.

Nigmatullin, C. M. 1976. Discovery of Giant Squid of the Genus *Architeuthis* in Atlantic Equatorial Waters. *Biol. Morya Vladivostok* 2(4):29–31.

Nishimura, S. 1966. Notes on the Occurrence and Biology of the Oceanic Squid, *Thysanoteuthis rhombus* Troschel in Japan. *Publ. Seto Mar. Biol. Lab.* 14(4): 327–49.

Nixon, M., and P. N. Dilly. 1977. Sucker Surfaces and Prey Capture. In M. Nixon and J. B. Messenger, eds., *The Biology of Cephalopods,* pp. 447–511. Academic Press.

Nordgård, O. 1923. The Cephalopoda Dibranchiata Observed Outside and in the Trondhjemsfjord. *K. norske Vidensk. Selsk. Skr.* 1922:1–14.

———. 1928. Faunistic Notes on Marine Evertebrates III. *K. norske Vidensk. Selsk. Forh.* 1(26):70–2.

Nordhoff, C. 1856. *Whaling and Fishing.* Moore, Wilsatsch, Keys & Co.

Norman, J. R., and F. C. Fraser. 1938. *Giant Fishes, Whales, and Dolphins.* W. W. Norton.

Norman, M. D., and C. C. Lu. 1997. Sex in Giant Squid. *Nature* 389:683–84.

Norris, K. S., and G. W. Harvey. 1972. A Theory for the Function of the Spermaceti Organ in the Sperm Whale (*Physeter catadon* L.). In S. R. Galler, K. Schmidt-Koenig, G. J. Jacobs, and R. E. Belleville, eds., *Animal Orientation and Navigation,* pp. 397–419. NASA.

Norris, K. S., and B. Møhl. 1983. Can Odontocetes Debilitate Prey with Sound? *American Naturalist* 122(1):85–104.

O'Dor, R. K. 1988a. The Energetic Limits on Squid Distributions. *Malacologia* 29(1):113–19.

———. 1988b. The Forces Acting on Swimming Squid. *Jour. Exp. Biol.* 137:421–42.

O'Dor, R. K., and R. E. Shadwick. 1989. Squid, the Olympian Cephalopods. *Jour. Ceph. Biol.* 1(1):33–55.

O'Dor, R. K., and D. M. Webber. 1986. The Constraints on Cephalopods: Why Squid Aren't Fish. *Canadian Jour. Zool.* 64:1591–605.

———. 1989. Invertebrate Athletes: Trade-Offs Between Transport Efficiency and Power Density in Cephalopod Evolution. *Jour. Exp. Biol.* 160:93–112.

O'Dor, R., H. O. Portner, and R. E. Shadwick. 1990. Squid as Elite Athletes: Locomotory, Respiratory, and Circulatory Integration. In D. L. Gilbert, W. J. Adelman, and J. M. Arnold, eds., *Squid as Experimental Animals,* pp. 481–503. Plenum.

Okutani, T. 1983. A New Species of an Oceanic Squid, *Moroteuthis pacifica* from the North Pacific (Cephalopoda; Onychoteuthidae). *Bull. Natn. Sci. Mus. Tokyo* (A)9:105–12.

————. 1995. *Cuttlefishes and Squids of the World in Color.* National Cooperative Association of Squid Processors Co.

Okutani, T., and T. Nemoto. 1964. Squids as the Food of Sperm Whales in the Bering Sea and Alaskan Gulf. *Sci. Rep. Whales Res. Inst.* 18:111–22.

Okutani, T., and Y. Satake. 1978. Squids in the Diet of 38 Sperm Whales Caught in the Pacific Waters off Northeastern Honshu, Japan, February, 1977. *Bull. Tokai Reg. Fish. Res. Lab.* 93:13–27.

Okutani, T., Y. Satake, S. Ohsumi, and T. Kawakami. 1976. Squids Eaten by Sperm Whales Caught off Joban District, Japan, During January-February. *Bull. Tokai Reg. Fish. Res. Lab.* 87:67–113.

Olafsen, E. 1772. *Vice-lovmand Eggert Olafsens og Lang-physici Biarne Povelson Reisen igiennen Island, foranstalter of Videnskabernes Saelskab i Kiobenhavn og beskreven of forbemeldte Eggert Olafsen.* Soro, J. Lindgrens Enke.

O'Shea, S. 1997. Giant Squid in New Zealand Waters. *Seafood New Zealand* 5 (10):32–34.

Owen, R. 1836. Descriptions of Some New or Rare Cephalopoda Collected by Mr. George Bennett. *Proc. Zool. Soc. London* 4:19–24.

————. 1880. Descriptions of Some New and Rare Cephalopoda. *Trans. Zool. Soc. London* 11:131–70.

Packard, A., H. E. Karlsen, and O. Sand. 1990. Low Frequency Hearing in Cephalopods. *Jour. Comp. Physiol.* 166:501–05.

Packard, A. S. 1873. Colossal Cuttlefishes. *American Naturalist* 7:87–94.

Pattie, B. H. 1968. Notes on the Giant Squid *Moroteuthis robusta* (Dall) Verrill Trawled off the Southwest Coast of Vancouver Island, Canada. *Wash. Dept. Fish. Res. Pap.* 3:47–50.

Pearcy, W. G. 1965. Species Composition and Distribution of Pelagic Cephalopods from the Pacific Ocean off Oregon. *Pacific Science* 19:261–66.

Pérez-Gándaras, G., and A. Guerra. 1978. Nueva cita de *Architeuthis* (Cephalopoda: Teuthoidea): Descripcion y alimentacion. *Inv. Pesq.* 42:401–14.

————. 1986. *Architeuthis* de Sudafrica: Nuevas citas y consideraciones biologicas. *Scient. Mar.* 53:113–16.

Peron, F. 1807. *Voyage de découvertes aux Terres Australes.* Paris.

Pfeffer, G. 1908. Die Cephalopoden. *Nordisches Plankton* 9:1–116.

————. 1912a. *The Cephalopoda of the Plankton Expedition*. English translation, 1992. Smithsonian Institution Libraries and National Science Foundation.

————. 1912b. Die Cephalopoden der Plankton-Expedition. Zugleich eine monographische Übersicht der oegopsiden Cephalopoden. *Ergebnisse der Plankton-Expedition der Humboldt-Siftung* 2:1–815.

Phillips, J. B. 1933. Description of a Giant Squid Taken at Monterey, with Notes on Other Squid Taken off the California Coast. *Calif. Fish and Game* 19(2):128–36.

————. 1961. Two Unusual Cephalopods Taken Near Monterey. *Calif. Fish and Game* 47(4):416–17.

Piatkowski, U., and W. Welsch. 1991. The Distribution of Pelagic Cephalopods in the Arabian Sea. *Bull. Mar. Sci.* 49(1-2):186–98.

Pickford, G. E. 1940. The Vampyromorpha, Living-Fossil Cephalopods. *Trans. N.Y. Acad. Sci.* 2(2):169–81.

————. 1946. *Vampyroteuthis infernalis* Chun. I. Natural History and Distribution. *Dana Report* 29:1–45.

————. 1949a. The Distribution of the Eggs of *Vampyroteuthis infernalis* Chun. *Sears Found. Jour. Mar. Res.* 8(1):73–83.

————. 1949b. *Vampyroteuthis infernalis* Chun. II. External Anatomy. *Dana Report* 32:1–131.

————. 1950. The Vampyromorphs (Cephalopoda) of the Bermuda Oceanographic Expeditions. *Zoologica* 35:87–95.

————. 1952. The Vampyromorpha of the "Discovery" Expeditions. *Discovery Reports* 26:197–210.

Pike, G. C. 1950. Stomach Contents of Whales off the Coast of British Columbia. *Prog. Rep. Pacific Cst. Stns.* 83:27–28.

Pinkas, L., M. S. Oliphant, and I. L. K. Iverson. 1971. Food Habits of Albacore, Bluefin Tuna, and Bonito in California Waters. *Calif. Dept. Fish and Game Fish. Bull.* 152:1–105.

Pippy, J. H. C., and F. A. Aldrich. 1969. *Hepatoxylon trichiuri* (Holden 1802) (Cestoda—Trypnorhyncha) from the Giant Squid *Architeuthis dux* Steenstrup 1857 in Newfoundland. *Canadian Jour. Zool.* 47(2):263–64.

Pliny. *Naturalis historia*. Loeb Classical Library, 1933. Harvard University Press.

Poli, F. 1958. *Sharks Are Caught at Night.* Regnery.

Polikoff, B. 1953. Unnatural Wonders of the World. *Chicago Nat. Hist. Bull.* 24(12):5.

Pontoppidan, E. 1755. *The Natural History of Norway.* London.

Poppe, G. T., and Y. Goto. 1993. *European Seashells.* Hemmen.

Posselt, H. J. 1891. *Todarodes sagittatus* (Lmk.) Stp. An Anatomical Study. With Remarks on the Relationships Between the Genera of the Ommastrephid Family. (1991 translation by J. Knudsen and M. Roeleveld.) *Steenstrupia* 17(5):161–96.

———. 1898. Grønlands Brachiopoder og Bløddyr. *Medd. om Grønland* 23:1–298 (269–83).

Proulx, E. A. 1993. *Shipping News.* Scribner's.

Quoy, J. R. C., and J. P. Gaimard. 1824. *Voyage autour du monde . . . exécuté sur les corvettes de S.M. l'Uranie et la physicienne pendant les années 1817, 1818, 1819, et 1820.* Louis de Freycinet. Paris.

Rae, B. B. 1950. Description of a Giant Squid Stranded near Aberdeen. *Proc. Malacol. Soc. London* 28:163–67.

Ramus, J. 1689. *Nori Regnum, hoc est Norvegica antiqua et ethnica, sive historiae Norwegicae prima initia a primo Norwegiae Rege, Noro, usque ad Haraldum Harfagerum,* etc. Christiana.

Rathjen, W. F. 1973. Northwest Atlantic Squids. *Mar. Fish. Rev.* 35(12):20–26.

Reed, D. C. 1995. *The Kraken.* Boyds Mills Press.

Rees, W. J. 1949. Giant Squid: The Quest for the Kraken. *Illustrated London News* 215:826.

———. 1950. On a Giant Squid *Ommastrephes caroli* Furtado Stranded at Looe, Cornwall. *Bull. Brit. Mus. Nat. Hist.* 1:31–41.

Rees, W. J., and G. E. Maul. 1956. The Cephalopoda of Madeira. *Bull. Brit. Mus. Nat. Hist.* 3:257–81.

Rice, D. W. 1978. Sperm Whales. In D. W. Haley, ed., *Marine Mammals of Eastern North Pacific and Arctic Waters,* pp. 82–87. Pacific Search Press.

Richard, J. N.d. *The Monaco Oceanographic Museum: Illustrated Guide.* Monte Carlo.

Ritchie, J. 1918. Occurrence of a Giant Squid *(Architeuthis)* on the Scottish Coast. *Scot. Naturalist* 133–39.

———. 1920. Giant Squid Cast Ashore at North Uist, Outer Hebrides. *Scot. Naturalist* 57.

———. 1922. Giant Squids on the Scottish Coast. *Rep. Brit. Assoc. Adv. Sci.* 1921:423.

Robison, B. H. 1989. Depth of Occurrence and Partial Chemical Composition of a Giant Squid, *Architeuthis,* off Southern California. *Veliger* 32(1):39–42.

———. 1995. Light in the Ocean's Midwater. *Scientific American* 273(1):60–64.

Robison, B. H., and R. E. Young. 1981. Bioluminescence in Pelagic Octopods. *Pacific Science* 35(1):39–44.

Robson, C. W. 1887. On a New Species of Giant Cuttlefish Stranded at Cape Campbell, June 30th, 1886. *(Architeuthis kirki). Trans. N.Z. Inst.* 20:34–39.

Robson, G. C. 1925. On *Mesonychoteuthis,* a New Genus of Oegopsid Cephalopoda. *Ann. Mag. Nat. Hist.* Series 9. 16:272–77.

———. 1929a. A Giant Squid from the North Sea. *Nat. Hist. Mag.* 2:6–8.

———. 1929b. *A Monograph of the Recent Cephalopoda.* Vol 1. London.

———. 1932. *A Monograph of the Recent Cephalopoda.* Vol. 2. London.

———. 1933. On *Architeuthis clarkei,* a New Species of Giant Squid, with Observations on the Genus. *Proc. Zool. Soc. London* 1933(3):681–97.

———. 1948. Cephalopoda Decapoda of the "Arcturus" Oceanographic Expedition, 1925. *Zoologica* 33:115–32.

Rodhouse, P. G., and M. R. Clarke. 1985. Growth and Distribution of Young *Mesonychoteuthis hamiltoni* Robson (Mollusca: Cephalopoda): An Antarctic Squid. *Vie Mileu* 35:23–30.

———. 1986. Distribution of the Early-Life Phase of the Antarctic Squid *Galiteuthis glacialis* in Relation to the Hydrology of the Southern Ocean in the Sector 15° to 30°E. *Mar. Biol.* 91:353–57.

Rodhouse, P. G., and E. M. C. Hatfield. 1990. Age Determinations in Squid Using Statolith Growth Increments. *Fish. Res.* 8:323–34.

Rodhouse, P. G., M. R. Clarke, and A. W. A. Murray. 1987. Cephalopod Prey of the Wandering Albatross *Diomeda exulans. Mar. Biol.* 96:1–10.

Rodhouse, P. G., P. A. Prince, M. R. Clarke, and A. W. A. Murray. 1990. Cephalopod Prey of the Grey-Headed Albatross *Diomeda chrysostoma*. *Mar. Biol.* 104:353–62.

Roeleveld, M., and J. Knudsen. 1980. Japetus Steenstrup: On the Merman (Called the Sea Monk) Caught in the Øresund in the Time of King Christian III. *Steenstrupia* 6(17):293–332.

Roeleveld, M. A. C., and M. R. Lipinski. 1991. The Giant Squid *Architeuthis* in Southern African Waters. *Jour. Zool. Soc. London* 224:431–77.

Roeleveld, M. A., C. J. Augustyn, and R. Melville-Smith. 1989. The Squid *Octopoteuthis* Photographed Live at 1,000 m. *S. Afr. Jour. Mar. Sci.* 8:367–68.

Rondelet, G. 1554. *Libri de piscibus marinis, in quibus verae piscium effigies expressae sunt.* Matthiam Bonhomme.

Roper, C. F. E. 1964. *Enoploteuthis anapsis,* a New Species of Enoploteuthid Squid (Cephalopoda: Oegopsida) from the Atlantic Ocean. *Bull. Mar. Sci. Gulf and Carib.* 14:140–48.

———. 1969. Systematics and Zoogeography of the Worldwide Bathypelagic Squid *Bathyteuthis. Bull. U.S. Nat. Mus.* 291:1–210.

———. 1977. Comparative Captures of Pelagic Cephalopods by Midwater Trawls. In M. Nixon and J. B. Messenger, eds., *The Biology of Cephalopods,* pp. 61–87. Academic Press.

———. 1983. An Overview of Cephalopod Systematics: Status, Problems and Recommendations. *Mem. Nat. Mus. Victoria* 44:13–27.

Roper, C. F. E., and K. J. Boss. 1982. The Giant Squid. *Scientific American* 246(4):96–105.

Roper, C. F. E., and M. Vecchione. 1993. A Geographic and Taxonomic Review of *Taningia danae* Joubin, 1931 (Cephalopoda: Octopoteuthidae) with New Records and Observations on Bioluminescence. In T. K. Okutani, R. K. O'Dor, and T. Kubodera, eds., *Recent Advances in Fisheries Biology,* pp. 441–56. Tokai University Press.

Roper, C. F. E., and G. L. Voss. 1983. Guidelines for Taxonomic Descriptions of Cephalopod Species. *Mem. Nat. Mus. Victoria* 44:49–63.

Roper, C. F. E., and R. E. Young. 1972. First Records of Juvenile Giant Squid *Architeuthis* (Cephalopoda: Oegopsida). *Proc. Zool. Soc. Wash.* 85(16):205–22.

———. 1975. Vertical Distribution of Pelagic Cephalopods. *Smithsonian Contributions to Knowledge* 209:1–51.

Roper, C. F. E., M. J. Sweeney, and C. F. Nauen. 1984. *FAO Species Catalogue. Vol. 3. Cephalopods of the World. An Annotated of Species of Interest to Fisheries.* Rome.

Roper, C. F. E., R. E. Young, and G. L. Voss. 1969. An Illustrated Key to the Families of the Order Teuthoidea. (Cephalopoda). *Smithsonian Contributions to Zoology* 13:1–32.

Saibil, H. R. 1990. Structure and Function of the Squid Eye. In D. L. Gilbert, W. J. Adelman, and J. M. Arnold, eds., *Squid as Experimental Animals,* pp. 371–98. Plenum.

Saidel, W. M., J. Y. Lettvin, and E. F. MacNichol. 1983. Processing of Polarized Light by Squid Photoreceptors. *Nature* 304:534–36.

Salcedo-Vargas, M. A. 1991. Checklist of the Cephalopods from the Gulf of Mexico. *Bull. Mar. Sci.* 49(1-2):216–20.

Salcedo-Vargas, M. A., and T. Okutani. 1994. New Classification of the Squid Family Magistoteuthidae (Cephalopoda: Oegopsida). *Venus (Jap.) Jour. Malacol.* 53(2):119–27.

Sanderson, I. 1956. *Follow the Whale.* Little, Brown.

Sasaki, M. 1929. A Monograph of the Dibranchiate Cephalopods of the Japanese and Adjacent Waters. *Jour. Fac. Agric. Hokkaido Inp. Univ.,* 20 (supp. 10):1–357.

Scammon, C. M. 1874. *The Marine Mammals of the North-Western Coast of North America, Together with an Account of the American Whale-Fishery.* Carmany and Company and G. P. Putnam's Sons.

Schaefer, F. S. 1992. Squids Live Fast, Die Young. *Sea Frontiers* 38(2):10–11.

Scheffer, V. B. 1969. *The Year of the Whale.* Scribner's.

Schlee, S. 1970. Prince Albert's Way of Catching Squid. *Natural History* 74(2):20–25.

Schlesinger, M. E., and N. Ramankutty. 1994. An Oscillation in the Global Climate System of Period 65–70 Years. *Nature* 367:723–26.

Searls, H. 1982. *Sounding.* Ballantine.

Sivertsen, E. 1955. Bleksprutt. *K. norske Vidensk. Selsk. Mus. Arb.* 1954:5–15.

Slijper, E. J. 1962. *Whales.* Basic Books.

Smith, A. G. 1963. More Giant Squids from California. *Calif. Fish and Game* 49(3):209–11.

Smith, A. M. 1991. Negative Pressure Generated by Octopus Suckers: A Study of the Tensile Strength of Water in Nature. *Jour. Exp. Biol.* 157:257–71.

———. 1996. Cephalopod Sucker Design and the Physical Limits to Negative Pressure. *Jour. Exp. Biol.* 199:949–58.

Solem, A. 1964. Giant Hunters of the Open Sea. *Chicago Nat. Hist. Mus. Bull.* 35(6):2–3.

Spaul, E. A. 1964. Deformity in the Lower Jaw of the Sperm Whale *(Physeter catadon). Proc. Zool. Soc. London* 142(3):391–95.

Starkey, J. D. 1963. I Saw a Sea Monster. *Animals* 2(23):629, 644.

Staub, F. 1993. Requin bleu, calmar géant et cachalot. *Proc. Roy. Soc. Arts and Sci. Mauritius* 5(3):141–45.

Stead, D. G. 1933. *Giants and Pygmies of the Deep: The Story of Australian Sea Denizens.* Sydney.

Steenstrup, J. 1849. Meddelese om tvende kiaempestore Blaeksprutter, opdrevne 1639 og 1790 ved Islands Kyst, og om nogle andre nordiskc Dyr. *Førh. skand. naturf.* 5:950–57.

———. 1855. Om den i Kong Christian IIIs tid Øresundet fange Havmand (Sømunken kaldet). *Dansk Maanedsskrift* 1:63–96.

———. 1857. Oplysninger om Atlanterhavets colossale Blaeksprutter. *Førh. skand. naturf.* 7:182–85.

———. 1898. Kolassale Blaeksprutter fra det nordlige Atlanterhav. *K. danske vidensk. Selsk. Skr.* (5):4(3):409–54.

———. 1849–1900. *The Cephalopod Papers of Japetus Steenstrup.* 1962 English translation by A. Volsoe, J. Knudsen, and W. Rees. Danish Science Press. Copenhagen.

Steinbach, H. B. 1951. The Squid. *Scientific American* 184(4):64–69.

Stendall, J. A. S. 1936. Giant Cuttlefish, *Sthenoteuthis caroli* Furtado, Ashore in Co. Londonderry. *Irish Nat. Jour.* 6:23–24.

Stephen, A. C. 1937. Notes on Scottish Cephalopods. *Scot. Naturalist* 131–32.

————. 1938. Rare Squid in Orkney. *Scot. Naturalist* 1938:119.

————. 1944. The Cephalopoda of Scottish and Adjacent Waters. *Trans. Roy. Soc. Edinburgh* 61:247–70.

————. 1950. Giant Squid, *Architeuthis* in Shetland. *Scot. Naturalist* 62(1):52–53.

————. 1962. The Species of *Architeuthis* Inhabiting the North Atlantic. *Proc. Roy. Soc. Edinburgh* 68:147–61.

Stephens, P. R., and J. Z. Young. 1978. Semicircular Canals in Squid. *Nature* 271:444–45.

Stevenson, J. A. 1935. The Cephalopods of the Yorkshire Coast. *Jour. Conch.* 20:102–16.

Storm, V. 1897. Om 2 udenfor Trondhjemsfojorden fundne kjaempeblaek-spruter. *Naturen* 97–102.

Straus, K. 1977. Jumbo Squid, *Dosidicus gigas. Oceans* 10(2):10–15.

Sweeney, J. B. 1972. *A Pictorial History of Sea Monsters and Other Dangerous Marine Life*. Crown.

Sweeney, M. J., C. F. E. Roper, K. M. Mangold, M. R. Clarke, and S. V. Boletzky, eds. 1992. "Larval" and Juvenile Cephalopods: A Manual for Their Identification. *Smithsonian Contributions to Zoology* 513.

Talmadge, R. C. 1967. Notes on Cephalopods from Northern California Waters. *Veliger* 10(2):200–02.

Tarasevich, M. N. 1968. The Diet of Sperm Whales in the North Pacific Ocean. *Zoologicheskiy Zhurnal* 47(4):595–601. (National Technical Information Service, 1974.)

Taylor, M. A. 1986. Stunning Whales and Deaf Squids. *Nature* 323:298–99.

Thiele, J. 1921. Die Cephalopoden der deutsch Sud-polar Expedition 1901–1903. *Dt. Sudpol. Exped.* 16(8):433–65.

Thomas, L. 1987. *Fantasi og virkelighed i naturen.* (Fantasy and Reality in Nature.) (Translated from the Danish by Mary Petersen.) Liberdan.

Thompson, D'A. W. 1901. On a Rare Cuttlefish, *Ancistroteuthis robusta* (Dall) Steenstrup. *Proc. Zool. Soc. London* 1900:992–98.

Thurston, H. 1989. Quest for the Kraken. *Equinox* 46:50–55.

Time-Life Books. 1988. *Mysterious Creatures.* Time Inc.

Toll, R. B., and S. C. Hess. 1981. A Small, Mature, Male *Architeuthis* (Cephalopoda: Oegopsida) with Remarks on Maturation in the Family. *Proc. Biol. Soc. Wash.* 94(3):753–60.

Topsell, E. 1607. *The Historie of Foure-Footed Beastes.* London.

Townsend, C. H. 1935. The Distribution of Certain Whales as Shown by Logbook Records of American Whaleships. *Zoologica* 29(1):1–50.

Tsuchiya, K., and T. Okutani. 1991. Growth Stages of *Moroteuthis robusta* (Verrill, 1881) with the Re-Evaluation of the Genus. *Bull. Mar. Sci.* 49(1-2):137–47.

———. 1993. Rare and Interesting Squids in Japan X. Recent Occurrences of Big Squids from Okinawa. *Venus* 52(4):299–311.

Turgeon, D. D., A. E. Bogan, E. V. Coan, W. K. Emerson, W. G. Lyons, W. L. Pratt, C. F. E. Roper, A. Scheltema, F. G. Thompson, and J. D. Williams. 1988. Common and Scientific Names of Aquatic Invertebrates from the United States and Canada: Mollusks. *American Fisheries Society Special Publication 16.*

Turner, H. J. 1963. Giant Squid. *Oceanus* 9(4):22–24.

Tweedie, M. 1963. Is There a Sea-Serpent? *Animals* 2(12):326–29.

Van Hyning, J. H., and A. R. Magill. 1964. Occurrence of the Giant Squid *(Moroteuthis robusta)* off Oregon. *Oregon Fish. Comm. Res. Briefs* 10:67–68.

Vecchione, M., and C. F. E. Roper. 1991. Cephalopods Observed from Submersibles in the Western North Atlantic. *Bull. Mar. Sci.* 49(1-2):433–45.

Vecchione, M., B. H. Robison, and C. F. E. Roper. 1992. A Tale of Two Species: Tail Morphology in Paralarval *Chiroteuhis* (Cephalopoda: Chiroteuthidae). *Proc. Biol. Soc. Wash.* 105(4):683–92.

Velain, C. 1875. Observations effectuées a l'Île Saint-Paul. *Compt. rendu. hebd.* 80:998–1003.

———. 1877. Remarques générales au sujet de la faune des Îles Saint-Paul et Amsterdam, suives d'une description de la faune malacologique de deux îles. *Archs. Zool. Exp. Gen.* 6:1–144.

Verne, J. 1870. *Twenty Thousand Leagues Under the Sea.* (1962 edition. Bantam.)

Verrill, A. E. 1874a. Occurrence of Gigantic Cuttle-Fishes on the Coast of Newfoundland. *Amer. Jour. Sci.* 7:158–61.

———. 1874b. The Giant Cuttle-Fishes of Newfoundland. *American Naturalist* 8:167–74.

———. 1875a. The Colossal Cephalopods of the Western Atlantic. *American Naturalist* 9:21–78.

———. 1875b. Notice of the Occurrence of Another Gigantic Cephalopod *(Architeuthis)* on the Coast of Newfoundland, in December 1874. *Amer. Jour. Sci.* 10:213–14.

———. 1876. Note on Gigantic Cephalopods—a Correction. *Amer. Jour. Sci.* 69:236–37.

———. 1877. Occurrence of Another Gigantic Cephalopod on the Coast of Newfoundland. *Amer. Jour. Sci.* 14:425–26.

———. 1879. The Cephalopods of the North-Eastern Coast of America. Part I: The Gigantic Squids *(Architeuthis)* and Their Allies; with Observations on Similar Large Species from Foreign Localities. *Trans. Conn. Acad. Sci.* 5:177–258.

———. 1880. The Cephalopods of the North-Eastern Coast of America. Part II: The Smaller Cephalopods, Including the "Squids" and the Octopi, with Other Allied Forms. *Trans. Conn. Acad. Sci.* 5:259–446.

———. 1881. Giant Squid *(Architeuthis)* Abundant in 1875, at the Grand Banks. *Amer. Jour. Sci.* 21:251–52.

———. 1882a. Occurrence of an Additional Specimen of *Architeuthis* at Newfoundland. *Amer. Jour. Sci.* 23:71–72.

———. 1882b. Some Remarks upon R. Owen's Paper on New and Rare Cephalopods. *Amer. Jour. Sci.* 23:72–75.

Verrill, G. E. 1958. *The Ancestry, Life and Work of Addison E. Verrill of Yale University.* Pacific Coast Publishing Co.

Villanueva, R. 1992. Deep-Sea Cephalopods of the North-Western Mediterranean: Indications of Up-Slope Ontogenetic Migration in Two Bathybenthic Species. *Jour. Zool. London* 227:267–76.

Villanueva, R., and A. Guerra. 1991. Food and Prey Detection in Two Deep-Sea

Cephalopods: *Opisthoteuthis agassizi* and *O. vossi* (Octopoda: Cirrata). *Bull. Mar. Sci.* 49(1-2):288–99.

Villanueva, R., and P. Sánchez. 1993. Cephalopods of the Benguela Current off Namibia: New Additions and Considerations on the Genus Lycoteuthis. *Jour. Nat. Hist.* 27(1):15–46.

Voss, G. L. 1953. A New Family, Genus and Species of Myopsid Squid from the Florida Keys. *Bull. Mar. Sci. Gulf and Carib.* 2(4):602–09.

———. 1956a. A Checklist of the Cephalopoda of Florida. *Q. Jour. Fla. Acad. Sci.* 19:274–82.

———. 1956b. A Review of the Cephalopods of the Gulf of Mexico. *Bull. Mar. Sci. Gulf and Carib.* 6:85–178.

———. 1959. Hunting Sea Monsters. *Sea Frontiers* 5(3):134–46.

———. 1960. Bermudan Cephalopods. *Fieldiana Zool.* 39(4):419–46.

———. 1962. South African Cephalopods. *Trans. R. Soc. S. Afr.* 36:245–72.

———. 1967. Squids: Jet-Powered Torpedoes of the Deep. *National Geographic* 131(3):386–411.

———. 1973. The Squid Boats Are Coming! *Sea Frontiers* 19(4):194–202.

———. 1977a. Classification of Recent Cephalopods. In M. Nixon and J. B. Messenger, eds., *The Biology of Cephalopods,* pp. 575–79. Academic Press.

———. 1977b. Present Status and New Trends in Cephalopod Systematics. In M. Nixon and J. B. Messenger, eds., *The Biology of Cephalopods,* pp. 49–60. Academic Press.

Voss, N. A. 1980. A Generic Review of the Cranchiidae (Cephalopoda; Oegopsida). *Bull. Mar. Sci.* 30(2):365–412.

Ward, R. 1948. *Henry A. Ward: Museum Builder to America.* Rochester Historical Society.

Ward, W. S. 1877. The New York Aquarium. *Scribner's Monthly* 13(5):577–91.

Webb, W. 1897. A Large Decapod. *Nautilus* 10:108.

Wells, H. G. 1905. "The Sea Raiders." In *Twenty-Eight Science Fiction Stories by H. G. Wells.* Scribner's. (Dover ed., 1952.)

Wells, M. J. 1964a. The Brain and Behavior of Cephalopods. In K. M. Wilbur and C. M. Yonge, eds., *Physiology of Mollusca,* vol. 2, pp. 547–90. Academic Press.

————. 1964b. Cephalopod Sense Organs. In K. M. Wilbur and C. M. Yonge, eds., *Physiology of Mollusca,* vol. 2, pp. 523–45. Academic Press.

Wells, M. J., and R. K. O'Dor. 1991. Jet Propulsion and the Evolution of Cephalopods. *Bull. Mar. Sci.* 49 (1-2):419–32.

Wendt, H. 1959. *Out of Noah's Ark.* Houghton Mifflin.

Whymper, F. 1883. *The Fisheries of the World. An Illustrated and Descriptive Record of the International Fisheries Exhibition, 1883.* Cassell & Company.

Williams, L. W. 1909. *The Anatomy of the Common Squid, Loligo pealei, Leseur.* E. J. Brill.

Wood, G. L. 1982. *The Guinness Book of Animal Facts and Feats.* Guinness Superlatives Ltd. London.

Wormuth, J. H. 1976. The Biogeography and Numerical Taxonomy of the Oegopsid Squid Family Ommastrephidae in the Pacific Ocean. *Bull. Scripps Inst. Oceanog.* 23:1–90.

Wray, J. W. 1939. *South Sea Vagabonds.* A. H. & A. W. Reed.

Young, J. Z. 1938a. The Giant Nerve Fibres and Epistellar Body of Cephalopods. *Q. Jour. Microsc. Sci.* 78:367–86.

————. 1938b. The Functioning of the Giant Nerve Fibres of the Squid. *Jour. Exp. Biol.* 15:170–85.

————. 1977. Brain, Behaviour and Evolution of Cephalopods. In M. Nixon and J. B. Messenger, eds., *The Biology of Cephalopods,* pp. 377–434. Academic Press.

————. 1984. The Statocysts of Cranchid Squids. *Jour. Zool.* 203:1–21.

Young, R. E. 1972. The Systematics and Areal Distribution of Pelagic Cephalopods from the Seas off Southern California. *Smithsonian Contributions to Zoology* 97:1–109.

————. 1975. Function of the Dimorphic Eyes in the Mid-Water Squid *Histioteuthis dofleini. Pacific Science* 29(2):211–18.

————. 1983. Oceanic Bioluminescence: An Overview of General Functions. *Bull. Mar. Sci.* 33(4):829–45.

————. 1995. Aspects of the Natural History of Pelagic Cephalopods of the Hawaiian Mesopelagic-Boundary Region. *Pacific Science* 49(2):143–55.

Young, R. E., and F. M. Mencher. 1980. Bioluminescence in Mesopelagic Squid: Diel Color Change During Counterillumination. *Science* 1286–88.

Young, R. E., and C. F. E. Roper. 1976. Bioluminescent Countershading in Midwater Animals: Evidence from Living Squid. *Science* 191:1046–48.

————. 1977. Intensity Regulation of Bioluminescence During Countershading in Living Midwater Animals. *Fish. Bull.* 75:239–52.

Young, R. E., C. F. E. Roper, and J. Walters. 1979. Eyes and Extra-Ocular Receptors in Midwater Cephalopods and Fishes: Their Role in Detecting Downwelling Light for Counterillumination. *Mar. Bio.* 51:371–80.

Young. R. E., E. M. Kampa, S. D. Maynard, F. M. Mencher, and C. F. E. Roper. 1980. Counterillumination and the Upper Depth Limit of Midwater Animals. *Deep-Sea Res.* 27A:671–91.

Yukhov, V. L. 1974. Records of the Giant Squids. *Priroda* 6:60–63.

Zeidler, W. 1981. A Giant Deep-Sea Squid, *Taningia* sp., from South Australian Waters. *Trans. Roy. Soc. S. Aust.* 105:218.

————. 1983. The Cephalopod Collection in the South Australian Museum. *Mem. Nat. Mus. Victoria* 44:77–79.

————. 1996. Forget the Calamari! *Australia Nature* 25(5):7–8.

Zeidler, W., and K. L. Gowlett-Holmes. 1996. A Specimen of Giant Squid, *Architeuthis* sp., from South Australian Waters. *Rec. S. Aust. Mus.* 29(1):85–91.

Illustration Credits

85   Henry Lee, *Sea Monsters Unmasked* (1884)

87   Henry Lee, *Sea Monsters Unmasked* (1884)

88   Verrill, *The Cephalopods of the North-Eastern Coast of America* (1879)

89   *Transactions of the Connecticut Academy of Sciences* 5: plate 54

93   Zoological Department, Museum of the University of Trondheim

95   Zoological Department, Museum of the University of Trondheim

99   Carla Skinder, New England Aquarium

100   *Adelaide Advertiser*

112   D. Paul, courtesy of Mark Norman

116   Courtesy of *New Zealand Herald*

134   Random House

135   Random House

136   *Illustrated London News,* 8 July 1875

139   Drawing by Glen Loates

143   Murray and Hjort, *The Depths of the Ocean* (1912)

147   Photograph by Alexander Remeslo

151   Drawing by Richard Ellis

155   Drawing by Glen Loates

161   Neg. No. 117165, courtesy of Department of Library Services, American Museum of Natural History

165   Photo: American Museum of Natural History

168   *Vingt mille lieues sous les mers*

169   Author's collection

171   *Vingt mille lieues sous les mers*

176   Author's collection

177   MJF Books

181   Frank Bullen, *The Cruise of the "Cachalot"*

188   Random House

193   Author's collection

195   Pierre Denys de Montfort, *Histoire naturelle générale et particulière des mollusques*

210   Poppe and Goto, *European Seashells*

217   Peabody Museum of Natural History, Yale University

219   Museum of Comparative Zoology, Harvard University

221   National Museum of Scotland

225   University of Rochester Library

Index

FOR THE BEST IN PAPERBACKS, LOOK FOR THE

In every corner of the world, on every subject under the sun, Penguin represents quality and variety—the very best in publishing today.

For complete information about books available from Penguin—including Puffins, Penguin Classics, and Arkana—and how to order them, write to us at the appropriate address below. Please note that for copyright reasons the selection of books varies from country to country.

In the United Kingdom: Please write to *Dept. EP, Penguin Books Ltd, Bath Road, Harmondsworth, West Drayton, Middlesex UB7 0DA.*

In the United States: Please write to *Penguin Putnam Inc., P.O. Box 12289 Dept. B, Newark, New Jersey 07101-5289* or call 1-800-788-6262.

In Canada: Please write to *Penguin Books Canada Ltd, 10 Alcorn Avenue, Suite 300, Toronto, Ontario M4V 3B2.*

In Australia: Please write to *Penguin Books Australia Ltd, P.O. Box 257, Ringwood, Victoria 3134.*

In New Zealand: Please write to *Penguin Books (NZ) Ltd, Private Bag 102902, North Shore Mail Centre, Auckland 10.*

In India: Please write to *Penguin Books India Pvt Ltd, 11 Panchsheel Shopping Centre, Panchsheel Park, New Delhi 110 017.*

In the Netherlands: Please write to *Penguin Books Netherlands bv, Postbus 3507, NL-1001 AH Amsterdam.*

In Germany: Please write to *Penguin Books Deutschland GmbH, Metzlerstrasse 26, 60594 Frankfurt am Main.*

In Spain: Please write to *Penguin Books S. A., Bravo Murillo 19, 1° B, 28015 Madrid.*

In Italy: Please write to *Penguin Italia s.r.l., Via Benedetto Croce 2, 20094 Corsico, Milano.*

In France: Please write to *Penguin France, Le Carré Wilson, 62 rue Benjamin Baillaud, 31500 Toulouse.*

In Japan: Please write to *Penguin Books Japan Ltd, Kaneko Building, 2-3-25 Koraku, Bunkyo-Ku, Tokyo 112.*

In South Africa: Please write to *Penguin Books South Africa (Pty) Ltd, Private Bag X14, Parkview, 2122 Johannesburg.*